Fabrizio Cei

Search for Neutrinos
from Stellar Gravitational Collapse
with the MACRO Experiment at Gran Sasso

TESI DI PERFEZIONAMENTO

SCUOLA NORMALE SUPERIORE
PISA - 1996

Tesi di perfezionamento in Fisica sostenuta il 20 dicembre 1995

COMMISSIONE GIUDICATRICE

Italo Mannelli, Presidente
Carlo Bemporad
Giuseppe Bertin
Samoil Bilenky
Francesco Fidecaro
Ettore Fiorini
Adalberto Giazotto

ISBN: 978-88-7642-284-3

Foreword

MACRO, with its 6 supermodules, is the largest liquid scintillation experiment in operation today. Although it has a lower mass than Čerenkov detectors, MACRO has a superior resolution in reconstructing event energy, position and timing.

From 1991 MACRO has in operation a supernova monitor for the on-line detection of neutrino bursts; this is based on dedicated circuits (*PHRASE*), acquisition system and analysis programs.

MACRO is therefore the first experiment to have studied the behaviour of such a device, over some years of data taking; the experiment is then in a position to alert astronomical observatories for an early measurement of the light emission by a supernova.

The list of the institutions involved in MACRO is:

- **Italy:**

    - Universities: Bari, Bologna, L' Aquila, Lecce, Napoli, Pisa, Roma La Sapienza, Torino;
    - Laboratori Nazionali di Frascati e del Gran Sasso.

- **U.S.:**

    - Universities: Boston, Drexel, Indiana, Michigan, Texas A & M;
    - California Institute of Technology.

The experiment used the financial, technical and logistic resources of the:

Istituto Nazionale di Fisica Nucleare (INFN)

Laboratori Nazionali del Gran Sasso (LNGS)

and of the:

U.S. Department of Energy (DOE).

Contents

List of Figures

List of Tables

Introduction

Neutrino Astrophysics is a new field of physics, which links together astronomy, astrophysics, cosmology, nuclear and particle physics; its birth and early development date back to the last ten years.

Low energy neutrino astrophysics studies the evolutionary life of the stars via the neutrinos emitted during the quiescent phase (*"solar neutrinos"*) and during the explosive death of big mass stars (*"supernova neutrinos"*); from the former, one obtains some information on the nuclear burning process in the star interior, from the latter, on the latest stage of the stellar evolution.

The neutrino mean free path in matter is ~ 20 orders of magnitude larger than that of the light; therefore, while the electromagnetic radiation reaching us comes from the external low-density layers of a star, ν's can be produced also in much deeper and higher-density layers. Since massless neutrinos are unaffected by their travel in the interstellar space, their energies and arrival directions carry information on the star history. On the other hand, many non-standard-model properties (mass, magnetic moment, lifetime, mixing ...) can be explored over distances and times ranging from $d \approx 1.5 \cdot 10^8$ Km and $t \approx 500$ s (neutrinos from the Sun) to $d \approx 1.5 \cdot 10^{18}$ Km and $t \approx 5 \cdot 10^{12}$ s (neutrinos from a supernova in the Large Magellanic Cloud, as SN1987A); the MSW effect can also be studied over a huge range of densities (from $\rho \sim 10^{-20} \, g \, cm^{-3}$ in the interstellar space to $\rho \sim 10^{14} \, g \, cm^{-3}$ in a supernova core).

The subject of this thesis is the search for neutrino bursts from galactic stellar gravitational collapses performed in the MACRO experiment, a large area modular detector, operating since autumn 1989 in a first partial configuration and since spring 1994 in its final configuration.

We shall discuss the stellar gravitational collapse phenomenology and its neutrino burst signature and briefly review the relevant present and future supernova neutrino detectors; then, we shall describe the MACRO experiment and the stellar gravitational collapse trigger; next, we shall examine the background sources, the background rejection techniques and the calibration methods; finally, we shall present the results of the search for neutrinos from stellar gravitational collapse and we shall discuss the MACRO potentialities in this search.

Chapter 1

Neutrinos from stellar gravitational collapses

1.1 Type I and Type II supernovæ

Stars have a complex evolution and their life is characterized by alternate stages of contraction and expansion.

First of all, the mutual gravitational attraction between the particles of the stellar plasma causes the star to contract; then, its central temperature increases and (if the core is hot enough) the Hydrogen nuclear fusion reaction is ignited. The nuclear burning converts four protons into a Helium nucleus and creates a thermal pressure, which counteracts the gravitational attraction and prevents the collapse of the stellar structure above the core. The star external layers expand, until an equilibrium condition is reached; our Sun, for instance, is burning its nuclear fuel in this quiescent phase since some billion years.

The proton fusion continues until the Hydrogen in the core is used up; then, the core contracts, because the gravitation is no longer opposed by the energy production, and both the core and the surrounding material are heated. Then, the Hydrogen burning begins in a shell surrounding the core and, if the core is hot enough, the Helium fusion reaction is ignited: three Helium nuclei are burned into a Carbon nucleus. This reaction also releases energy and causes the thermal pressure. Further reactions can take place, burning Carbon to form Oxygen, Neon and Silicon.

The star develops an "onion-skin" structure, with a central core of Silicon and Sulfur, surrounded by shells of Neon, Oxygen, Carbon and Helium and an outer envelope of Hydrogen. The fusion last cycle combines Silicon nuclei to form Iron; ^{56}Fe is the final stage for spontaneous fusion, because it is the most strongly bound nucleus and a further ^{56}Fe fusion would absorb, rather than release, energy.

Only big mass stars ($M \gtrsim 10\ M_\odot$) can supply a sufficient gravitational energy to burn Helium, Carbon etc., while in lower mass stars only a part of the nuclear fusion chain can take place.

Different categories of stars can experience, in the latest stages of their life, a sudden, many orders of magnitude, increase of luminosity; such phenomenon is known as *"supernova explosion"* ([BE90b], [WO86a]). Astronomers distinguish two types of supernovæ, Type I and Type II, depending on the presence or absence of Hydrogen lines in their spectra: Type II supernovæ have strong Hydrogen lines, Type I have none.

Type I supernovæ are usually believed [BE90b] to be due to the thermonuclear disruption of white dwarfs, which are parts of a binary system and accrete their core mass from a companion star, at a rate $\dot{M} \sim 10^{-(5 \div 10)} \ M_\odot/year$. When the mass of the nucleus grows to the Chandrasekhar limit ($M_{Ch} \approx 5.8 \, Y_e^2 \ M_\odot \approx 1.4 \ M_\odot$)* ($Y_e$ is the electron fraction, i.e. the number of electrons per baryon), the structure collapses and Carbon (or Helium) is ignited in highly degenerate conditions. The final product of this burning process is ^{56}Fe, which derives from the β-decay of ^{56}Ni into ^{56}Co and of ^{56}Co into ^{56}Fe; this process causes the emission of a small amount of low-energy, practically undetectable, neutrinos. The absence of the Hydrogen lines is naturally explained, because all the Hydrogen accreted from the companion star is quickly converted into Helium.

Type I supernovæ are further divided in two subclasses:

- $\sim 10\,\%$ of the supernovæ without Hydrogen lines have peculiar spectra, without Silicon absorption [WH90]. Silicon is expected in the white dwarf thermonuclear disruption as an intermediate stage of the Iron production; its absence would imply a quick $Si \rightarrow Fe$ conversion, which is unlikely in this scheme. Supernovæ without Hydrogen and Silicon lines might be massive stars which lost their Hydrogen envelope during the supergiant stage. These supernovæ are called Type Ib and explode, as that of Type II, by a different mechanism, the gravitational collapse of their core;

- the other Type I supernovæ are called Type Ia.

Some further distinctions can be introduced, but they are unimportant in this context and will be omitted.

The mechanism of Type Ib and Type II supernovæ is essentially understood, but many details are still uncertain.

The basic idea ([BE90b], [WO86a], [BU88], [BU90b], [WI93], [JA93]) is that a big mass ($M \gtrsim 10 \ M_\odot$) star experiences a fast core collapse, followed by the formation of a shock wave which propagates through the star and, after reaching its surface, propels the stellar material into space. The collapse originates from the formation, by ^{28}Si burning, of a degenerate ^{56}Fe core, at a central temperature $T_c \sim 5 \times 10^9 \ ^\circ K$ and a central density $\rho_c \sim 10^{10} \ g \, cm^{-3}$. As

---

*The Chandrasekhar limit is the maximum allowed mass for a hydrodynamically stable structure, supported by the electron degeneracy pressure.

already stated, Iron burning cannot lead to any heavier element, Iron being the most strongly bound nucleus; Iron can be photodisintegrated, absorbing energy and further accelerating the star contraction, so that the burning net result is an increase in mass, temperature and density of the Iron core, which rapidly reaches the Chandrasekhar limit. The infall of the stellar material continues for ~ 1 s, until the central density grows up to $\rho_c \approx 5 \times 10^{14}\, g\,cm^{-3}$ (a few times higher than the nuclear density), when the nuclear pressure becomes dominant, slowing down and finally stopping the collapse and producing a "bounce" of the core[‡].

During this infall phase, the extremely high density makes the electron capture by protons

$$e^- + p \rightarrow n + \nu_e \qquad (1.1.1)$$

very efficient; a first ν_e burst, with a total energy $E_{inf} \sim 10^{51}$ erg, is therefore expected ("*infall burst*"). However, the electron capture does not continue indefinitely because, at a density $\rho_{trap} \approx 10^{12}\, g\,cm^{-3}$, ν_e's are trapped, i.e. their mean free path ($\lambda \sim 0.5$ Km) is much smaller than the radius of the core ($R \sim 30$ Km). The ν_e's go into equilibrium with the electrons via the inverse process

$$\nu_e + n \rightarrow e^- + p \qquad (1.1.2)$$

and the "*lepton fraction*" $Y_l = Y_e + Y_{\nu_e}$ (the lepton number per baryon) of the core is frozen at $Y_l \approx 0.33 \div 0.40$.

The core "bounce" produces sound waves moving outward, while the infalling material is travelling inward; the infall velocity (V) and the local sound velocity (A), as a function of the radius of the star (in Km), are shown in fig. (1.1). The curves cross at a radius $R_{cros} \approx 25$ Km; their crossing point is called the "*sonic point*". At the sonic point, the sound velocity in the star reference frame is zero; then, the sound waves cannot get beyond it and a shock wave forms.

The scheme outlined above is generally accepted but, after the formation of the shock, the mechanism is not so clear. Two basic ideas have been proposed: the prompt-shock model ([CO60] et al.) and the delayed-shock model ([WI85] et al.; for a review see for instance [JA93]).

The prompt-shock model predicts that the shock will go through the whole star and expel most of it, producing a supernova and leaving behind the unshocked core, which becomes a neutron star; the large gravitational energy of the neutron star is released essentially as neutrinos. This simple scenario is not confirmed by the results of most numerical simulations, in which the shock loses energy, dissociating nuclei into nucleons, and finally stalls, at a radius $R_{stal} \sim 400$ Km; therefore, no more outward motion is possible and the prompt shock fails to expel the outer part of the star. The success of the

---

[‡]Also lower mass stars ($8\, M_\odot \lesssim M \lesssim 10\, M_\odot$) sometimes explode as supernovæ: in this case, the nuclear burning stops with the formation of a degenerate O, Ne and Mg core, which collapses after accreting to the Chandrasekhar limit.

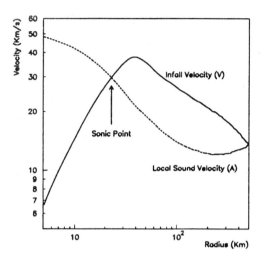

Figure 1.1: Infall (V) and local sound velocity (A) as a function of the star radius. The curve crossing point is the *"sonic point"* (adapted from [AR77]).

prompt-shock model requires a small Iron core (less the material which must be traversed, less the shock energy loss); since the mass of the Iron core increases with the initial mass of the star, the prompt-shock model is likely only for intermediate mass stars: $M \lesssim 10\ M_\odot$.

The delayed-shock model, on the other hand, predicts that the stalled shock can be revived, after some hundred ms from the bounce, by the neutrinos emitted by the hot neutron star. Neutrinos diffuse in the mantle (the region outside the unshocked core) and can dissociate nuclei into nucleons, heating the material and reviving the shock, which will then expel the star material. Neutrinos are assumed, in a good approximation, to originate from a *"neutrinosphere"* of radius R_ν, defined as the region from which the neutrinos can freely escape to infinity. At this stage, neutrinos of all flavours (e, μ, τ) are present, produced by the pair annihilation process

$$e^- + e^+ \rightarrow \nu_x + \bar\nu_x \qquad (1.1.3)$$

While ν_e and $\bar\nu_e$ can interact and lose energy via neutral and charged current reactions, $\nu_{\mu,\tau}$ and $\bar\nu_{\mu,\tau}$ can interact only via neutral currents, because their energy is below the μ production threshold; therefore, their mean energy is higher than that of the other neutrinos and their neutrinosphere is deeper within the core. The neutronization process makes the supernova mantle a neutron-rich environment; therefore, ν_e charged current interactions ($\nu_e + n \rightarrow e^- + p$) are more probable than $\bar{\nu}_e$ charged current interactions ($\bar{\nu}_e + p \rightarrow e^+ + n$) and $\langle E_{\nu_e} \rangle < \langle E_{\bar{\nu}_e} \rangle$.

The effects of the various neutrino processes on the material reheating were discussed by many authors ([MY90], [HA88a], [HA88b], [BR91]); an important contribution looks to be due to neutral current interactions of higher energy ν_μ's and ν_τ's, greatly enhanced by the giant resonance excitation [HA88a].

Up to the nineties, most supernova numerical simulations were one-dimensional, i.e. they assumed a spherically symmetrical collapse, neglecting the rotation of the star around its axis, the turbulence and the convective motions of the envelope etc. The reason for that was essentially the necessity of a reasonable computer time; however, two and three-dimensional simulations (e.g. [JA94], [BU95]) recently became a practical possibility thanks to faster new-generation computers. In the multi-dimensional models the neutrino heating mechanism is more effective than in the one-dimensional models, so that the supernovæ explode more easily.

The delayed-shock mechanism is generally successful for a wide range of star masses and is practically the unique effective mechanism for very big mass stars ($M \gtrsim 20\ M_\odot$); also the explosion of *SN1987A* has been explained by some authors (e.g. [BR87]) using a delayed-shock mechanism. As an example of the results of this mechanism, the trajectories of various mass points are shown in fig. (1.2) (from [WI85]). The inner dashed curve is the electron neutrinosphere, the outer dashed curve represents the shock and the 1.665 $M_\odot$ is the lowest mass point which is propelled outward; the mass increases from bottom to top. The radius of the star is given in *cm* and the time in *s* ($t = 0$ at the bounce).

When the shock crosses the electron neutrinosphere at a radius $R_{\nu_e} \sim$ 80 Km, the efficiency of the process (1.1.1) suddenly increases, because of the large number of shock-liberated free protons; a second sharp ν_e burst (*"neutronization"* or *"prompt breakout burst"*), which deleptonizes the outer core and forms a proto-neutron star, is therefore emitted, $\sim 5 \div 10$ ms after the bounce. The total energy of this second burst is $\lesssim 10^{52}\ erg$ and its duration ~ 10 ms. Also the other neutrinospheres are crossed by the shock, causing the copious ν_x, $\bar{\nu}_x$ pairs (1.1.3) generated in the shock-heated mantle to escape, but the neutrino luminosity in this stage is dominated by the ν_e burst.

The unshocked core of the star cools slowly ($t_{cool} \sim 20$ s) by ν_x, $\bar{\nu}_x$ diffusion and emission; the residual degenerates to a neutron star (or a black hole), with a final lepton fraction $Y_l \approx Y_e \approx 0.04$, a mass $M_{NS} \approx M_{Ch}$ and a radius $R_{NS} \approx 10$ Km (fig. (1.3), after [BU90b]).¶

¶Some questions about the mechanism of the collapse are still open, expecially for what concerns the strong interaction physics; for instance, the equation of state of the extremely degenerate nuclear matter is poorly known and the effects of the nuclear incompressibility on the neutrino diffusion timescale must be studied more carefully.

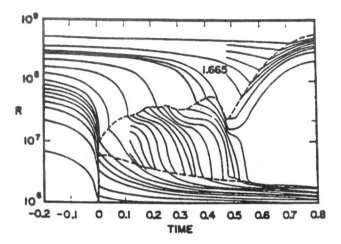

Figure 1.2: Trajectories of various mass points in the Wilson delayed-shock model [WI85]. The radius of the star is given in *cm* and the time in *s*; $t = 0$ at the bounce.

The collapse makes the core more tightly gravitationally bound, causing the release of an extra energy

$$\Delta E = -G \left(\frac{M_{ST}^2}{R_{ST}} - \frac{M_{NS}^2}{R_{NS}} \right) \approx 2 \div 4 \times 10^{53} \; \text{erg} \approx 0.1 \div 0.2 \; M_\odot \qquad (1.1.4)$$

where G is the Newton gravitational constant and M_{ST} and R_{ST} are the mass and radius of the uncollapsed star. The Iron photodisintegration energy and the kinetic energy of the expelled material can be simply estimated [MO91]; both contributions are at maximum a few per cent of the total ($\sim 2 \div 3 \times 10^{51}$ erg). The energy of the optical flare-up is much smaller ($\lesssim 10^{49}$ erg), so it is clear that the bulk ($\approx 99\%$) of the gravitational binding energy is carried away by the cooling neutrinos (1.1.3).

Also some gravitational radiation could be emitted; since the symmetry of the collapse is largely spherical, the mass distribution quadrupole moment is small and the amount of the energy released in gravitational waves represents a negligible fraction ($\sim 10^{-5}\%$) of the total ([BU94], [PI90]). However, the detection of these gravitational waves by dedicated experiments would have a great importance, gravitational waves being the most direct test of the general relativity hypothesis. A more comprehensive picture of the stellar collapse mechanism would emerge if neutrinos and gravitational waves were simultaneously detected; the neutrinos are emitted from the core outer layers,

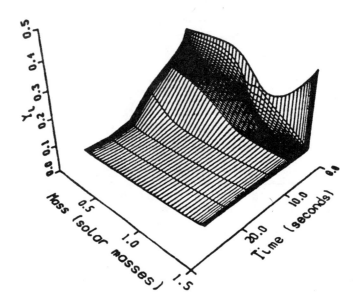

Figure 1.3: The lepton fraction $Y_l = Y_e + Y_{\nu_e}$ per baryon versus baryon mass (in units of $M_\odot$) and time (in s) for a 1.4 $M_\odot$ proto-neutron star model. The neutron star values $Y_l \approx 0.04$ is reached after ≈ 20 s (after [BU90b]).

while the gravitational waves come from the core center, providing an observational information on the infalling matter motion. Moreover, a correlated neutrino-gravitational wave detection was recently suggested as a possible tool for measuring a ν_e mass in a few eV range [SW94]: for a 1 eV ν_e mass, the infall burst would be recorded with a ≈ 0.25 ms delay in the respect of the gravitational wave pulse. An accuracy of a fraction of ms in the absolute time measurement looks within the reach of interferometric gravitational wave detectors (like *VIRGO* and *LIGO*, Laser Interferometer Gravitational Wave Observatory) and stellar collapse neutrino experiments.

(The most important sources of gravitational radiation are probably the coalescing neutron binary stars ([PI89], [PI92], [PI93]). A collision between two neutron stars would produce a modulated amplitude, almost monochromatic, gravitational wave signal; the strength of this signal at the Earth would be an order of magnitude larger than that emitted by a type II supernova at the same distance. These collisions would also be efficient sources of γ-ray (with an energy release $E \approx 10^{51}$ erg) and neutrino bursts (energy release $\sim 10^{53}$ erg). The interest of a time correlated observation of gravitational wave and γ-ray bursts was stressed by Piran [PI92]; in principle also a simultaneous observation of gravitational waves and neutrino bursts would be possible for these systems. The binary merging frequency is $\sim 1/$ galaxy $/10^{4 \div 5}$ years; if all the galaxies up to the Virgo Cluster (distance ≈ 10 Mpc) are considered, this corresponds to $\sim 1/year$. Such a rate would make the simultaneous detection a realistic possibility for experiments having a ~ 10 year live time; unfortunately, while gravitational wave detectors and γ-ray observatories are sensitive to such distant phenomena, the present (and probably future) neutrino burst detectors are not. They are limited to galactic neutrino bursts; then, the neutrino-gravitational wave simultaneous detection is practically precluded by the too low rate of the binary coalescence in the Milky Way.)

1.2   Energy and time distributions of neutrinos from stellar collapse

The features of the neutrino burst (neutrino flux, total energy, energy spectrum, luminosity curve, neutrino flavour composition ...) carry information on the star evolution during the gravitational collapse.

As already discussed, three phases (*infall, neutronization* and *cooling*) can be distinguished in the neutrino emission; some authors do not distinguish between the first two stages, since the neutrino-producing process is, in both cases, the electron capture (1.1.1). The total energy released in the first two phases is $\sim 10^{52}$ *erg*; the corresponding neutrino fluxes are so low that present neutrino experiments (and maybe also some future experiments) have no chance to detect the infall burst and a small chance to detect the neutronization burst. The infall and neutronization neutrino energy distributions have the structure of capture spectra, with a mean value $\bar{E} \approx 10$ MeV.

The cooling phase is clearly more interesting, since neutrinos carry away $\approx 99\%$ of the energy emitted during the collapse. The features of the cooling burst were calculated by many authors; here some models are briefly reviewed.

A classical model [NA80] is the one by Nadezhin and Otroshchenko, who calculated the neutrino energy spectra at two different "epochs": immediately after the collapse ($t = 0.04$ s) and when the hot neutron star has been formed ($t = 5.1$ s in their calculation). The energy spectrum at the second epoch can be approximated by a pseudo Fermi-Dirac distribution, with a high-energy deficit

$$\Phi(E_\nu) = A \frac{(E_\nu/T)^2}{1 + \exp(E_\nu/T)} \exp\left[-\alpha(E_\nu/T)^2\right] \qquad (1.2.1)$$

where E_ν is the neutrino energy, T is the temperature of the neutrinosphere, the parameter α takes into account the partial non transmission of the star layers above the neutrinosphere and the normalization constant A is determined by the total energy released in neutrinos. The values of the constant α, of the total energy E_{tot} and of the temperature T for the various kinds of neutrinos are listed in table (1.1).

Table 1.1: Parameters of the neutrino energy spectra in the [NA80] model.

Neutrinos	E_{tot} (erg)	T (MeV)	α
ν_e	1.1×10^{53}	3.5	0.01
$\bar{\nu}_e$	10^{53}	4.5	0.02
$\nu_{\mu,\tau}, \bar{\nu}_{\mu,\tau}$	10^{53}	8	0

The qualitative picture of the collapse in the Nadezhin-Otroschenko model is similar to that of more recent calculations, but the total energy release (6.1×10^{53} erg) is about twice larger; therefore the number of events in a detector predicted by this model should be regarded as rather optimistic.

Bruenn and Haxton ([BR91], [BU90b] and references therein) calculated the neutrino energy spectra at the first stages of the collapse, including the inelastic neutrino-nuclei interactions, which provide an efficient shock reheating mechanism. These interactions soften the energy spectra, which are then "less than thermal" and can be approximated by:

$$\Phi(E_\nu) = A \frac{(E_\nu/T)^2}{1 + \exp\left(\frac{E_\nu - \mu}{T}\right)} \qquad (1.2.2)$$

where μ is the neutrino chemical potential. The ratio between μ and T is indicated by η; typical values for T and η are 3.4, 4.2 and 6 MeV and 1.8, 2 and 2.8 for ν_e, $\bar{\nu}_e$ and $\nu_{\mu,\tau}$ ($\bar{\nu}_{\mu,\tau}$). The normalization constant A is chosen by comparing this spectrum with a pure Fermi-Dirac distribution ($\mu = 0$) having the same mean energy and the same number of neutrinos. The energy spectra (1.2.2) are shown in fig. (1.4); ν_x indicates ν_μ and ν_τ and their antineutrinos.

Similar results were obtained by Myra and Burrows [MY90], whose model was used by Burrows et al. for calculating the expected neutrino signal in the present and future detectors [BU92]. In this model the energy spectra have a "pinched" shape, i.e. they show a deficit of both low and high energy neutrinos with respect to thermal distributions with zero chemical potential[§]. The total energy E_{tot}, the mean energy $\bar{E}$ and the parameter η ($= \mu/T$) of these spectra are listed in table (1.2); ν_x indicates the sum of $\nu_{\mu,\tau}$ and $\bar{\nu}_{\mu,\tau}$ (the contribution to the flux of each of them is one fourth of the total).

[§]The high-energy deficit is a common feature of many supernova neutrino models; on the other hand, some calculations show high-energy tails; see for instance [MA87a].

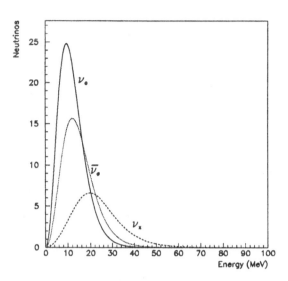

Figure 1.4: Neutrino and antineutrino energy spectra in Bruenn and Haxton calculation (adapted after [BR91]); the neutrinos are in units of $10^{55}/\,\mathrm{MeV}$.

A summary of the results of many neutrino spectra calculations can be found in [JA93]; in table (1.3) we show the mean energy, the parameter η and the fraction f of the total energy carried away by each neutrino flavour as quoted in this paper. The mean energies must be red-shifted by $10 \div 20\,\%$ for an observer at infinity to take into account the motion of the neutron star surface.

The evolution of the neutrino luminosity in the [MY90] model in the first 230 ms after the onset of the collapse $(t = 0)$ is shown in fig. (1.5); the bounce occurs at $t \approx 110$ ms. The general features of the neutrino signal can be easily observed: a ν_e flash at the shock break-out, an abrupt turn-on of $\bar{\nu}_e$ and "ν_μ"$(= \nu_{\mu,\tau}, \bar{\nu}_{\mu,\tau})$, an initial rise of the luminosity after the shock, followed by a slower, approximately exponential, decrease, lasting ~ 10 s. The ν_e, $\bar{\nu}_e$ luminosity oscillations are due to hydrodynamic pulsations of the mantle.

An "experimental support" to these theoretical calculations has been provided by the $SN1987A$ explosion; the small number of observed events (see later) prevented the investigation of the model details and simple thermal distributions (including or not a neutrinosphere cooling) were used to fit the $\bar{\nu}_e$ energy spectra.

As an example: Bludman and Schinder [BL88] used a Fermi-Dirac spectrum

Table 1.2: Parameters of the neutrino energy spectra in [BU92] model.

Neutrinos	E_{tot} (erg)	$\bar{E}$ (MeV)	η
ν_e	5.87×10^{52}	9.9	1.2
$\bar{\nu}_e$	5.18×10^{52}	11.6	2.0
ν_x	1.79×10^{53}	15.4	3.0

Table 1.3: Typical values of the neutrino energy spectra parameters; the quoted uncertainties are from [JA93] paper.

Neutrinos	$\bar{E}$ (MeV)	η	Energy fraction f
ν_e	$10 \div 12$	$3 \div 5$	$0.17 \div 0.22$
$\bar{\nu}_e$	$14 \div 17$	$2 \div 2.25$	$0.17 \div 0.28$
ν_x	$24 \div 27$	$0 \div 2$	$0.5 \div 0.66$

with zero chemical potential in two different hypotheses:
  a) constant $\bar{\nu}_e$ temperature;
  b) neutrinosphere cooling, with a negative power law

$$T(t) = \frac{T_0}{\left(1 + \frac{a t}{n}\right)^n} \qquad (1.2.3)$$

(the latter gives the best likelihood value). The constants a and n and the initial temperature T_0 which fit the energy spectrum and the time structure of the neutrino burst are respectively: $a = 0.14 \pm 0.05$ s^{-1}, $n = 0.4$ and $T_0 = 4.2 \pm 0.5$ MeV. For the constant temperature model $T = 3.3$ MeV gives the best fit; the total radiated energy is $E_{tot} = 3.4 \times 10^{53}$ erg in both cases.

Using this model in calculations, the usual assumptions are: a) the neutrino temperatures are $T_{\nu_e} = T_{\bar{\nu}_e} = 3.3$ MeV (constant temperature model) or $T_0(\nu_e) = T_0(\bar{\nu}_e) = 4.2$ MeV (cooling model), and $T_{\nu_\mu} = T_{\nu_\tau} = T_{\bar{\nu}_\mu} = T_{\bar{\nu}_\tau} = 6.6$ MeV (constant temperature model) or $T_0(\nu_\mu) = T_0(\nu_\tau) = T_0(\bar{\nu}_\mu) = T_0(\bar{\nu}_\tau) = 8.4$ MeV (cooling model); b) the released energy is equally shared between all the neutrino species. Because the total neutrino energy is proportional to the fourth power of the temperature and the neutrino flux to the third power, the ν_e, $\bar{\nu}_e$ flux is twice the $\nu_{\mu,\tau}$, $\bar{\nu}_{\mu,\tau}$ flux. The [BL88] constant temperature energy spectra are shown in fig. (1.6).

Other authors (e.g. [BA87a]) used a Boltzmann ($dN/dE \sim E^2 \exp{(-E/T)}$) instead of a Fermi-Dirac spectrum; the inferred $\bar{\nu}_e$ temperature and total energy ($T_{\bar{\nu}_e} \approx 4$ MeV, $E_{\bar{\nu}_e} \approx 5 \times 10^{52}$ erg) are in general agreement with [BL88].

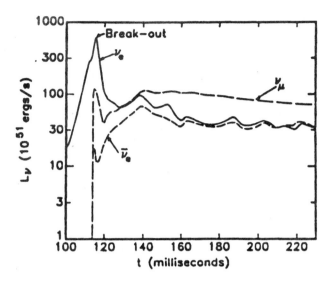

Figure 1.5: Neutrino luminosity as a function of time; ν_μ indicates muon and tau neutrinos and antineutrinos collectively (after [MY90]).

1.3   Neutrino flavour mixing in a supernova core

The neutrino spectra discussed in the previous section were calculated under the hypothesis of standard-model (i.e. massless) neutrinos. On the other hand, a group of important hints (for instance solar ([GA92a], [GA94a], [KA90], [KA91a], [DA93]) and atmospheric ([KA92], [IM92]) neutrino data and the apparent need for a mixed non-baryonic Dark Matter [CA94a]) suggest the possibility of (at least) one non-zero neutrino mass and neutrino oscillations ([PO68], [BI78]). The solar neutrino puzzle, if interpreted in terms of MSW conversion ([MI86], [WO78]), singles out two possible regions in the $\left(\delta_m^2, \sin^2\left(2\,\theta_V\right)\right)$ plane ($\delta_m^2 = m_{\nu_1}^2 - m_{\nu_2}^2$, ν_1 and ν_2 being two neutrino mass eigenstates, and θ_V is the vacuum mixing angle): the "large angle" ($\delta_m^2 \approx 10^{-(5\div7)}\,eV^2$ and $\sin^2\left(2\,\theta_V\right) \approx 0.6 \div 0.9$) and the "small angle" ($\delta_m^2 \approx 3 \times 10^{-6} \div 10^{-5}\,eV^2$ and $\sin^2\left(2\,\theta_V\right) \approx 10^{-(2\div3)}$) solution ([BE91], [GA92b], [GA94a], [BL92], [BA93b], [BE93]). The atmospheric ν_μ deficit could be explained by $\nu_\mu \leftrightarrow \nu_e$ oscillations with parameters: $\delta_m^2 \approx 10^{-(2\div3)}\,eV^2$ and

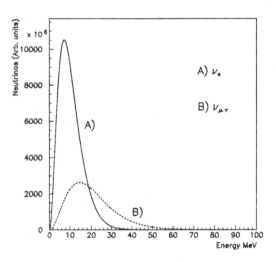

Figure 1.6: ν_e and $\nu_{\mu,\tau}$ energy spectra in [BL88] constant temperature model.

$\sin^2(2\theta_V) \approx 0.5 \div 0.9$ [BE92] or by $\nu_\mu \leftrightarrow \nu_x$ ($x = \tau$ or sterile) oscillations with parameters: $\delta_m^2 \approx 10^{-(1\div2)}\,eV^2$ and $\sin^2(2\theta_V) \gtrsim 0.5$ [GA94b]. The nucleosynthesis theory requires that $> 90\%$ of the Dark Matter to be in a non baryonic form and the structure of the universe, over a very large range of distance scales, favours a mix of cold and hot not baryonic Dark Matter [CA94a]; a plausible candidate for the hot Dark Matter is one or more species of neutrinos, having a total mass in the range $1 \div 100$ eV ([CO72], [BO80]).

MSW conversion (at the moment, the most plausible explanation of the solar neutrino deficit (for instance [CA94b], [BE93])) is strongly dependent on the density of the traversed medium; then, it is likely that large effects are associated with this mechanism inside a supernova core.

The MSW mechanism is based on the observation that electron neutrinos interact in matter in a different way than μ and τ neutrinos. The reason is the same which is responsible for the large difference between ν_e and ν_μ (or ν_τ) elastic scattering cross section on electrons: the contribution of the charged current diagram to the matrix element of the $\nu_e - e$ interaction. In an electron sea of number density $N_e = \rho\, y_e$ (ρ is the medium mass density and y_e is the number of electrons per baryon), a ν_e acquires an effective mass squared

$$m_{eff}^2 \approx \frac{4\,G_F}{\sqrt{2}}\,\rho\, y_e\, E_{\nu_e} \qquad (1.3.1)$$

where G_F is the Fermi constant and E_{ν_e} is the ν_e energy. This is the classical Wolfenstein result [WO78]. In a supernova core, at least one other effect must be taken into account: the contribution of the ν_e neutral current exchange diagram to the neutrino-neutrino scattering amplitude [FU87]. Neglecting higher order contributions and inserting appropriate units, the ν_e effective mass squared can be expressed by:

$$m_{eff}^2 \left(eV^2 \right) \approx 1.52 \times 10^3 \, E_{\nu_e} \, (\mathrm{MeV}) \, \rho_{10} \, (y_e + y_{\nu_e}) \qquad (1.3.2)$$

where ρ_{10} is the mass density in units of $10^{10} \, g \, cm^{-3}$ and y_{ν_e} is the number of electron neutrinos per baryon. If neutrinos (at least one) are massive, the leptonic number is not absolutely conserved and the Hamiltonian operator is not diagonal in the flavour basis. Now consider a neutrino created as a ν_e at $t = 0$; at a time $t > 0$ its wavefunction $|\nu(t)>$ will be a superimposition of different flavour eigenstates. Therefore, there will be a non-zero conversion probability $\nu_e \to \nu_x$ given by:

$$P\left(\nu_e \to \nu_x\right) = |<\nu_x|\nu(t)>|^2 = \sin^2\left(2\,\theta_M\right)\sin^2\left(\frac{\pi\,c\,t}{L_M}\right) \qquad (1.3.3)$$

where θ_M is the neutrino mixing angle in matter, c is the speed of light and L_M is the matter oscillation length. θ_M and L_M are related to m_{eff}^2, $\delta_m^2 = m_{\nu_x}^2 - m_{\nu_e}^2 \approx m_{\nu_x}^2$ and the vacuum mixing angle θ_V by the equations

$$\sin^2\left(2\,\theta_M\right) = \frac{\delta_m^4 \sin^2\left(2\,\theta_V\right)}{\left(m_{eff}^2 - \delta_m^2 \cos\left(2\,\theta_V\right)\right)^2 + \delta_m^4 \sin^2\left(2\,\theta_V\right)} \qquad (1.3.4)$$

$$L_M = \frac{4\,\pi\,E_\nu}{\sqrt{\left[\left(m_{eff}^2 - \delta_m^2 \cos\left(2\,\theta_V\right)\right)^2 + \delta_m^4 \sin^2\left(2\,\theta_V\right)\right]}} \qquad (1.3.5)$$

In (1.3.4) there is a resonance condition which maximizes the oscillation amplitude $(2\,\theta_M = \pi/2)$; the corresponding oscillation length is

$$L_{res} \approx \frac{4\pi\,E_\nu}{\delta_m^2 \sin\left(2\,\theta_V\right)} \approx \frac{126\,E_\nu\,(\mathrm{MeV})}{\delta_m^2\,(eV^2)\,\theta_V} \quad cm \qquad (1.3.6)$$

where the last approximation assumes $\theta_V \ll 1$.

A plausible (but without any definitive evidence !) scenario which accommodates both a $\nu_e \leftrightarrow \nu_\mu$ level crossing in the Sun and a cosmologically significant neutrino mass is obtained choosing $m_{\nu_\mu} \sim 10^{-3}$ eV and $m_{\nu_\tau} \sim 1 \div 100$ eV. Given these values, a sequential level crossing occurs in the decreasing density supernova layers: a $\nu_e \leftrightarrow \nu_\tau$ mixing in the mantle, followed by a $\nu_e \leftrightarrow \nu_\mu$ mixing in the presupernova hydrogen envelope (see fig. (1.7)). The effects of such masses on neutrino fluxes and spectra depend on the vacuum mixing angles $\theta_{e,\mu}$ and $\theta_{e,\tau}$: for the former, the central value suggested by the solar

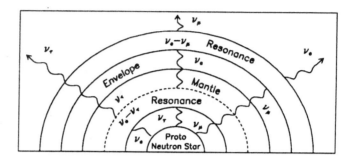

Figure 1.7: Schematic picture of neutrino flavour mixing in a supernova core
(after [QI94])

neutrino puzzle ($\sin^2 (2\,\theta_{e,\mu}) \approx 4 \times 10^{-6}/\delta m_{e,\mu}^2 \approx 4 \times 10^{-6}/m_{\nu_\mu}^2$) is the most
obvious choice, while for the latter there are no direct suggestions. However,
an interesting constraint can be set on $\sin^2 (2\,\theta_{e,\tau})$ considering how the heavy
element nucleosynthesis in supernovæ would be modified by a too large mixing
angle ([QI93], [QI95]).

Nucleosynthesis requires a neutron-rich environment; the equilibrium between neutrons and protons in the supernova ejecta is maintained by the charged current reactions

$$\nu_e + \text{n} \rightarrow \text{e}^- + \text{p} \tag{1.3.7}$$

$$\bar{\nu}_e + \text{p} \rightarrow \text{e}^+ + \text{n} \tag{1.3.8}$$

A large $\nu_e \leftrightarrow \nu_x$ mixing would produce many high-energy electron neutrinos, enhancing the (1.3.7) efficiency. In the MSW mechanism the $\bar{\nu}_e$'s do not
undergo oscillations (unless $m_{\nu_e} > m_{\nu_x}$); then, the efficiency of the neutron
production reaction (1.3.8) would not be enhanced; the net result should be a
reduction of the neutron amount. With a cosmologically significant neutrino
mass ($m_{\nu_x} \sim 100$ eV), a mixing angle $\sin^2 (2\,\theta_{e,\tau}) > 10^{-5}$ would produce such
a large reduction of the neutron/proton ratio to preclude the possibility of an
efficient nucleosynthesis. A lower ν_x mass ($m_{\nu_x} \sim 10$ eV) causes this upper
bound to decrease: $\sin^2 (2\,\theta_{e,\tau}) \lesssim 10^{-4}$.

The general effect of a $\nu_e \leftrightarrow \nu_x$ mixing would be a hardening of the ν_e
spectrum in respect to the no-mixing case. This can be particularly important
for the neutrino heating, because the (1.3.7) reaction (whose cross section rises
with the square of the neutrino energy) is responsible of a large energy transfer

from the neutrinos to the shock; in case of mixing, up to a 60 % energy transfer increase would be expected [FU92].

1.4   Expected number of events in a detector

The computation of the expected number N_{ev} of neutrino events in a detector is straightforward. For a chosen reaction, N_{ev} is given by (for simplicity we do not distinguish between neutrinos and antineutrinos):

$$N_{ev} = \frac{1}{4\pi R^2} \int_0^\infty dt \int_0^\infty dE\, \Phi_{T,\alpha}(E,t)\, \epsilon(E)\, \sigma(E)\, N_{tar} \qquad (1.4.1)$$

where t is the time, E is the neutrino energy, R is the supernova distance from the Sun, $\Phi_{T,\alpha}(E,t)$ is the neutrino spectrum (T is the neutrino temperature and α indicates a set of unspecified parameters, as the chemical potential μ), N_{tar} is the number of the target nuclei, $\sigma(E)$ is the cross section and $\epsilon(E)$ is the detector efficiency.

The neutrino spectrum can be written in the form

$$\Phi_{T,\alpha}(E,t) = N_\nu\, f_{T,\alpha}(E,t) \qquad (1.4.2)$$

where N_ν is the neutrino total number and $f_{T,\alpha}(E,t)$ is the normalized spectrum

$$\int_0^\infty dt \int_0^\infty dE\, f_{T,\alpha}(E,t) = 1 \qquad (1.4.3)$$

Therefore, inserting (1.4.3) into (1.4.1) we have

$$N_{ev} = \frac{N_\nu\, N_{tar}}{4\pi R^2} \int_0^\infty dt \int_0^\infty dE\, f_{T,\alpha}(E,t)\, \epsilon(E)\, \sigma(E) \qquad (1.4.4)$$

The integration in (1.4.4) can sometimes be performed analytically, but it is usually performed numerically. An alternative and often useful form of (1.4.4) is obtained assuming, for simplicity, that the efficiency ϵ can be taken outside of the integration and replaced by an "overall" efficiency $\bar{\epsilon}$. Now, at the right hand side we are left with the integration of the cross section $\sigma(E)$ over the neutrino normalized spectrum $f_{T,\alpha}(E,t)$; performing this integration is equivalent to calculate the average value $\langle\sigma\rangle_{T,\alpha}$ of the cross section on the neutrino spectrum.

Then (1.4.4) becomes

$$N_{ev} = N_\nu\, N_{tar}\, \bar{\epsilon}\left(\frac{\langle\sigma\rangle_{T,\alpha}}{4\pi R^2}\right) \qquad (1.4.5)$$

If, for instance, we consider $\bar{\nu}_e$ reactions on protons in liquid scintillator detectors, we have (using the constant temperature model [BL88] and assuming $E_{th} \approx 7$ MeV)

$$N_{ev} \simeq 22000\left(\frac{M_{det}(kton)}{R^2(Kpc)}\right) \qquad (1.4.6)$$

Therefore, from a supernova at $10\,\mathrm{Kpc} \approx 220$ events are expected in a 1-kton liquid scintillation detector. The crude estimate (1.4.6) shows that a stellar collapse neutrino detector must be very massive ($M_{det} \sim 1$ kton). More detailed calculations of the expected number of events in operating and proposed detectors will be given later.

1.5 *SN1987A*

On February 23 1987, $7^h\,35^m$ Universal Time ($U.T.$) the first non-optical observation of a supernova ($SN1987A$) was performed by two underground water Čerenkov detectors, *Kamiokande II* [KA87] and *IMB*3 (Irvine Michigan Brookhaven, [IM87]), primarily sensitive to $\bar{\nu}_e$ capture on protons: $\bar{\nu}_e + p \rightarrow$ e^+ + n. The supernova was optically visible, for the first time, on a photographic plate taken on February 23, $10^h\,40^m$ $U.T.$ [MA87b]; it was not visible in a previous observation at $9^h\,2^m$ $U.T.$ The supernova progenitor was a 18 $M_\odot$ blue-giant star (*Sanduleak 69202*), located at $R.A. = 5^h\,35^m\,49.992^s$ and $Decl. = -69°\,17'\,50.08''$ in the Large Magellanic Cloud, at a distance $R \approx 50\,\mathrm{Kpc}$ from the Sun.

Looking at (1.4.5) and (1.4.6) one easily observes that the expected number of neutrino events from such a distant stellar collapse is quite small, $\approx 9\ \bar{\nu}_e$ events in a 1-kton liquid scintillation detector; in the case of the water Čerenkov detectors this number is reduced to $\approx 7/kton$. Both Kamiokande II and IMB3 were nucleon decay detectors; their characteristics are reported in table (1.4).

Table 1.4: Characteristics of Kamiokande II and IMB3 detectors.

	Kamiokande II	IMB3
Mass (tonn)	3000	8000
Fiducial Volume (tonn)	2140	6800
Number of PMT's	948	2048
PMT diameter (inches)	20	8
PMT coverage (%)	20	5
Energy resolution $\sigma(E)/E$	$20\%/\sqrt{E/(10\ \mathrm{MeV})}$	$35\%/\sqrt{E/(10\ \mathrm{MeV})}$
Angular resolution (degrees)	27 %	$\sim 15 \div 20\%$
	($E = 10$ MeV)	($E \approx 20 \div 40$ MeV)
Energy threshold (MeV)	7	20

These experiments had very different efficiencies at low energies: because of the better ratio between the surface covered by the PMT's and the total surface (corresponding to a better collection efficiency for the Čerenkov light) and the lower background, Kamiokande II had a sharper efficiency curve (which reached 90 % at 15 MeV) and a lower energy threshold[†] than IMB3 (see last line of (1.4) table). The IMB3 efficiency curve reached 90 % at $\approx$ 50 MeV; therefore it was sensitive only to the high energy tail of the neutrino spectrum and recorded 8 events, to be compared with 12[*] events recorded by Kamiokande II, which had a smaller sensitive mass.

Two liquid scintillation detectors, *Baksan* in the North Caucasus and *LSD* (Liquid Scintillation Detector) in the Mont Blanc tunnel, also reported signals which were attributed to neutrinos from *SN1987A*.

Baksan, a 200 *tonn* neutrino telescope, reported a 6 event burst (one event was later identified as a muon), whose energies span from 12 to 23 *MeV*, in 9 *s*; Baksan burst occurred $\approx$ 25 s after the first IMB3 event [BA87b]. This time discrepancy could be reconciled by an uncertainty of $\approx$ 50 *s* in the Baksan absolute clock [BA88c], so that this burst is generally believed to be associated with *SN1987A*. A 5 event signal in 200 tonn of liquid scintillator is about a factor 3 larger than what expected from (1.4.6); it can be explained invoking the large statistical fluctuations of the small numbers and some uncertainties in the background rejection.

LSD, a 90 *tonn* galactic supernova neutrino detector, reported a 5 event burst in 8 *s* [LS87], 4.7 hours before the Kamiokande/IMB burst. The measured energies of these events are between 7 and 11 *MeV*, close to the detector threshold. This signal is controversial, but the general opinion is that it cannot be associated with *SN1987A* for various reasons: not only the time discrepancy is very difficult to overcome in any astrophysical plausible scenario (for an attempt, see for instance [DR87]), but the LSD burst is by itself "non-standard". In fact, only in a scenario with a very low electron antineutrino temperature ($T_{\bar{\nu}_e} \approx 1\ MeV$) and an extremely high total energy release ($E_{tot} \approx 10^{55}\ erg$, about 2 orders of magnitude larger than the binding energy of a neutron star) a 5 event burst in LSD without a copious burst in Kamiokande II (which had a slightly higher energy threshold) would be expected [SC88]. Some people [AG89a] reported gravitational wave detector noise in Italy and in Maryland in coincidence with the LSD burst (similar "coincidences" were found also for Kamiokande II and IMB3 signals [AG91]). However, these antennas are room temperature, i.e. intrinsically noisy, detectors; a significant signal would imply a total energy $E_{GW} > 2000\ M_{\odot}$ emitted in gravitational waves [SC88]. A tentative explanation of the Mt. Blanc events was suggested by Haxton [HA87]: while water Čerenkov detectors are largely insensitive to neutral current reactions, a significant number of neutral current events can be induced in liquid scintillator detectors by $\nu_{\mu,\tau}$, $\bar{\nu}_{\mu,\tau}$ exciting the 15.1 MeV ^{12}C level (see later

---

[†]Some months before the supernova explosion, Kamiokande II had been upgraded in order to detect solar neutrinos via elastic scattering on electrons.

[*]The sixth Kamiokande II event does not match the Kamioka people criteria for a good event and is usually excluded from the data analysis.

(2.1.9)). A $\bar{\nu}_e$-poor (and $\nu_{\mu,\tau}$-rich) burst could have caused the Mt. Blanc signal, without a corresponding signal in Kamiokande. This interpretation does not require extreme assumptions on $\nu_{\mu,\tau}$ temperature, but does not solve the problem of the 4.7-hour difference between the two bursts.

Returning to the more established Kamiokande II and IMB3 neutrino burst, it must be remarked that the number of observed events in both detectors is in reasonable agreement with theoretical predictions, taking into account their different sensitivities; moreover, comparing these numbers with the Wilson and Mayle calculations [MA87a], one obtains indications for a progenitor mass $M_{pr} \sim 15\ M_\odot$, consistent with the optical determination of the *Sanduleak 69202* mass.

The observed durations of the neutrino burst are 12.4 s in Kamiokande II and 5.6 s in IMB3; the fact that the burst durations are ~ 10 s rather than tens of ms is an indication of the neutrino diffusion out of a very dense core (cooling stage).

It must be observed that the Kamiokande II clock had a poor Universal Time accuracy ($\approx 1^m$), while the accuracy of the IMB3 clock was much better (≈ 50 ms); therefore a time correlated analysis of the events from the two detectors can be performed only if one makes some assumptions about the relative synchronization of the clocks. Assuming that the first neutrinos arrived simultaneously in Kamiokande II and IMB3, one obtains the energy vs time distribution shown in fig. (1.8).

The Kamiokande II event time distribution shows a 7.3 s gap: 9 events were clustered in the first 2 s, while the remaining 3 events occurred 7.3 s after the ninth event and within 3.2 s. This time gap was discussed by many authors ([SC88], [BA89a], [BL88]), but no particular significance can be attributed to it. Looking at fig. (1.8) it can be noted that two IMB3 events fall in the Kamiokande II gap, assuming, as previously stated, that the first events in the two detectors were simultaneous.

The angular distributions of the Kamiokande II and IMB3 events (fig. (1.9)) are consistent with an isotropic distribution (as expected from $\bar{\nu}_e + p \rightarrow e^+ + n$ events), even if in the Kamiokande II data a small excess in the LMC direction is present. Again, the low statistics prevents from taking this excess too seriously. In principle, it could be due to one or two electron scattering events, which have a forward-peaked angular distribution; in this case, supernova models with high energy tails are favoured, because pure thermal distributions predict less than 0.5 scattering events for *SN1987A*. However, the most conservative interpretation is that this excess is due to a fluctuation [SC88].

Kamiokande II and IMB3 data can be compared in an attractive way [SC88] looking at the plot in fig. (1.10), which shows the emitted energy in $\bar{\nu}_e$, $\epsilon_{\bar{\nu}_e}$, and the total emitted energy, ϵ_T, versus the electron antineutrino temperature $T_{\bar{\nu}_e}$ for both experiments; statistical (1 σ) and systematic errors are taken into account in drawing the contour lines. The regions corre-

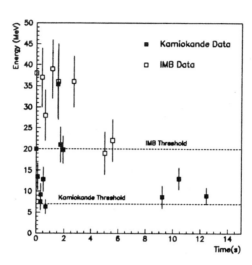

Figure 1.8: Energy vs time distribution of the Kamiokande II (filled squares) and IMB3 (open squares) data, obtained assuming that the first neutrinos arrived simultaneously on both detectors (adapted after [BÖ92]).

sponding to the two experiments overlap close to the standard model values ($\epsilon_T \approx 3 \times 10^{53}$ erg, $T_{\bar{\nu}_e} \approx 4.5$ MeV)[§].

1.5.1   Constraints on neutrino properties from *SN1987A* observations

Constraints on neutrino properties were obtained from *SN1987A* observations using the facts that all the events are clustered within 12 s, their energies are lower than 40 MeV and the neutrinos practically account for all the energy released during the collapse ([MO91], [BA89a], [BÖ92], [SC88], [SC90], [AR89]).[**]

a) *Neutrino mass.* A finite $\bar{\nu}_e$ mass would cause a neutrino signal time spreading; more precisely, two massive neutrinos emitted at the same time

---

[§]In this picture the exponent "53" of $\epsilon_{\bar{\nu}_e}$ scale is wrong and must be replaced by "52".
[**]See also the references quoted in these books and papers.

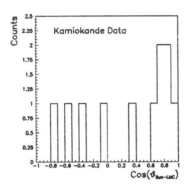 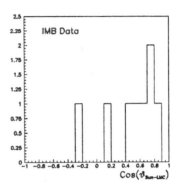

Figure 1.9: Angular distribution of the Kamiokande II (left) and IMB3 events (right). The sixth Kamiokande event was excluded.

would arrive on the Earth with a time separation

$$
\Delta t_a = \frac{D}{c}\left(\frac{1}{\beta_1} - \frac{1}{\beta_2}\right) \approx \frac{D}{c}\left(\frac{1}{1 - \frac{m_\nu^2}{2E_1^2}} - \frac{1}{1 - \frac{m_\nu^2}{2E_2^2}}\right) \approx
$$
$$
\approx \frac{D}{c}\left(1 + \frac{m_\nu^2}{2E_1^2} - 1 - \frac{m_\nu^2}{2E_2^2}\right) = \frac{D}{2c}m_\nu^2\left(\frac{1}{E_1^2} - \frac{1}{E_2^2}\right) \quad (1.5.1)
$$

where $D \approx 50$ Kpc is the SN1987A distance, m_ν is the neutrino mass (in units of energy), c is the speed of light and E_1 and E_2 are the neutrino energies; for $m_\nu = 10$ eV, this corresponds to a separation $\Delta t_a \approx 2.5$ s between a 10 MeV and a 50 MeV neutrinos. However, the intrinsic time spread Δt_e of the burst must be taken into account; assuming $\Delta t_e \leq 10$ s and performing a statistical analysis of the Kamiokande II and IMB3 data, limits in the range $19 \div 30$ eV were set on the $\bar{\nu}_e$ mass. Model-dependent calculations lower this limit to 11 eV. The limits are not far from the terrestrial tritium β decay limit: $m_{\bar{\nu}_e} \leq 4.8$ eV [BE94].

b) $\bar{\nu}_e$ charge. A non-zero neutrino electric charge Q_ν would also produce a neutrino signal time broadening, because neutrinos with different energies would have different paths in the Galactic Magnetic Field; higher energy neutrinos would move along a straighter path and arrive on the Earth before lower energy neutrinos [BA87c]. Using a statistical analysis the following limit

$$
\frac{Q_\nu}{e} \leq 3 \times 10^{-17}\left(\frac{1}{B(\mu G)\, x_G(\text{Kpc})}\right) \quad (1.5.2)
$$

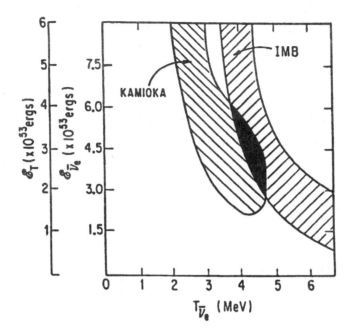

Figure 1.10: Emitted energy $\epsilon_{\bar{\nu}_e}$ in $\bar{\nu}_e$ and total emitted energy ϵ_T versus $\bar{\nu}_e$ temperature for Kamiokande II and IMB3 data, allowing for statistical (1 σ) and systematic errors (after [SC88]).

was set. In this formula $B\,(\sim 2 \div 3\ \mu G)$ is the Galactic Magnetic Field and x_G is the path length in the field.

c) *Neutrino lifetime.* The fact that the ν's travelled a distance $D \approx 50\,\mathrm{Kpc}$ without decaying sets a limit on the $\bar{\nu}_e$ lifetime at rest $\tau_{\bar{\nu}_e}$

$$\gamma \tau_{\bar{\nu}_e} \gtrsim 1.6 \ \times 10^5 \ years \qquad (1.5.3)$$

where γ is the relativistic factor E_ν / m_ν. For a 10 MeV $\bar{\nu}_e$ this corresponds to

$$\tau_{\bar{\nu}_e} \gtrsim 5 \times 10^5 \ m_{\bar{\nu}_e}(\mathrm{eV})\ s \qquad (1.5.4)$$

and rules out ν-decay as a solution of the solar neutrino puzzle, unless one combines neutrino decay with some assumptions on neutrino mixing: a short lifetime cannot be excluded in models with large mixing angles for an unstable neutrino state having a large ($\sim 50\%$) ν_e component [FR88].

Limits on the radiative neutrino decay were also set, comparing the experimental limits on the photon flux with that expected from the decay of unstable

ν_e, $\nu_{\mu,\tau}$ bursts [FE88]

$$\frac{\tau_{\bar{\nu}_e}}{m_{\bar{\nu}_e}} \geq 8.3 \times 10^{14} \text{ s eV}^{-1} \tag{1.5.5}$$

$$\frac{\tau_{\nu_{\mu,\tau}}}{m_{\nu_{\mu,\tau}}} \geq 3.3 \times 10^{14} \text{ s eV}^{-1} \tag{1.5.6}$$

d) *Neutrino magnetic moment and right-handed coupling.* A large $\bar{\nu}_e$ magnetic moment would led to spin-flip interactions within the star. On the basis of the existing observations a limit

$$\mu_{\bar{\nu}_e} \lesssim 10^{-12} \; \mu_B \tag{1.5.7}$$

(μ_B is the Bohr magneton) was obtained [BA88b]. Also the right-handed weak coupling strength G_{RH} can be constrained by using the cooling time scale of *SN1987A*

$$\frac{G_{RH}}{G_F} \lesssim 10^{-4} \tag{1.5.8}$$

where G_F is the Fermi constant [RA88].

e) *Limiting velocity and weak equivalence principle.* The special and general relativity hypotheses can be tested by using the difference in arrival times between neutrinos and photons (less than 3 hours).

First of all, the speed of photons and neutrinos cannot differ by more than 1 part in 10^8 [ST88], which is an accurate check of the special relativity hypothesis that all objects have the same limiting velocity.

Following the special relativity, the speed of each massless particle must be an universal constant, independent on the motion of the particle source. If we write

$$c'_\nu = c_\nu + K_\nu v_s \tag{1.5.9}$$

for the velocity of neutrinos emitted by a moving source (with velocity v_s), the ≤ 3 *hour* time difference between neutrinos and photons sets on K_ν the limit [AT94]

$$K_\nu \leq 10^{-5} \tag{1.5.10}$$

A less stringent limit ($K_\nu \leq 2 \times 10^{-4}$) follows from the broadening of the neutrino pulse induced by the (1.5.9) relation [AT94].

Secondly, assuming that the difference in arrival time between ν's and γ's is caused solely by the passage through the galactic gravitational field, Longo [LO88] and Krauss and Tremaine [KR88] obtained that photons and neutrinos move along the same trajectory to an accuracy of at least 0.5%. This is a test of the weak equivalence principle, which states that all types of radiation follow the same path in a gravitational field. Note that the statement is highly conservative, since it neglects the propagation time of the shock-wave to the stellar surface (it takes ~ 1 *hour*, see ([SH87], [AR88])) and the rise time of the supernova luminosity up to observable magnitudes.

f) *Neutrino mixing angle and $\nu_{\mu,\tau}$ masses.* *SN1987A* data provided the possibility to set an upper bound on the neutrino mixing angle and on the $\nu_{\mu,\tau}$ masses.

A large $\nu_e \leftrightarrow \nu_x$ conversion would produce a hardening of the $\bar{\nu}_e$ spectrum; so, the cumulative energy spectrum (i.e. the fraction $F\left(E < \bar{E}\right)$ of the total energy observed in a detector for event energies $E < \bar{E}$) would drift towards higher energies. The experimental cumulative energy spectra rule out a neutrino mixing angle $sin^2\left(2\,\theta\right) \geq 0.7 \div 0.9$ at 90 % C. L. Note that this mixing angle range includes the "large angle" solution of the solar neutrino problem and a part of the region of the $\nu_e \leftrightarrow \nu_\mu$ oscillation space which could be responsible for the atmospheric ν_μ deficit [SM94].

If μ and/or τ neutrinos are massive, they can experience helicity flip processes via neutral current interactions on nucleons, becoming almost sterile. In this case, the neutron star cooling would be rapidly accelerated and the time distribution of the emitted neutrino burst would be distorted and shortened. Kamiokande II and IMB3 event time distributions were used by Gandhi and Burrows [GA90] to set a limit on $\nu_{\mu,\tau}$ masses: $m_{\nu_{\mu,\tau}} = \sqrt{m_{\nu_\mu}^2 + m_{\nu_\tau}^2} \leq 28$ KeV (there is no way to distinguish ν_μ and ν_τ in these calculations).

Probably the most important conclusion to be drawn from *SN1987A* is that neutrino astronomy is a practical possibility, even with neutrino detectors not specifically designed for this purpose, and that many interesting results about the supernova theory and neutrino properties could be obtained if neutrinos from a galactic supernova would be observed.

1.6   Galactic supernova rate

How many supernovæ explode in a century in our Galaxy is a fundamental question; unfortunately, the answer is rather controversial.

The present neutrino detectors are sensitive only to galactic supernovæ (recently some ideas for extragalactic supernova neutrino detectors were proposed ([CL90], [CL94], [KO92]); these projects will be discussed later); they have good chances of success only if the average time interval between two consecutive galactic supernova explosions is at maximum of the order of magnitude of their life time, which is ~ 10 years. If, on the contrary, the expected number of galactic stellar collapses is of order of one per century or less, a completely different experimental approach is required: one can only search for relic neutrinos from past supernovæ, without much hope to detect a "real-time" collapse. Different calculations of this number give results in the range $1 \div 10$ stellar gravitational collapses per century, which is not so high to fully justify the construction of dedicated neutrino detectors and at the same time is not so low to make the real-time stellar collapse neutrino detectors useless. Actually, all the present stellar collapse neutrino detectors were built looking at other physics subjects also: magnetic monopoles, neutrino astronomy, neutrino oscillations, solar neutrinos, proton decay, cosmic rays etc.

The most direct method to evaluate the rate of the galactic supernova explosions is based on a sky long-time monitoring by means of telescopes, CCD images, photographic plates etc. The idea is to measure the occurrence rate of supernovæ in other galaxies and to infer the rate in the Milky Way by using an appropriate normalization, based on the galactic luminosity. Since the absolute luminosity of a galaxy is proportional to the inverse square of its distance from the Sun and therefore to the square of the Hubble constant H_0, the uncertainty (about a factor 2) on H_0 produces a corresponding uncertainty on the galactic supernova rate. This technique is also biased by the fact that a stellar gravitational collapse not necessarily produces a bright supernova explosion: in many models, copious neutrino bursts are emitted together with a modest electromagnetic energy release; moreover, weak supernovæ are difficult to detect and can be obscured by the cosmic dust and gas. Finally, inclination effects must be taken into account, because supernovæ are more easily discovered in galaxies with viewing angles $i \lesssim 30°$.

An evaluation of the number N_{Gal} of Type Ib and II galactic supernovæ per century was performed by Van den Bergh and Tammann [VA91a], using the results of some systematic surveys

$$N_{Gal} = 7.3 \ h^2 \qquad\qquad (1.6.1)$$

where h is the Hubble constant in units of $100 \ Km \ s^{-1} \ Mpc^{-1}$. In a more recent paper [VA93] Van den Bergh re-examined the applied viewing corrections and found them too large; the new estimated rate of galactic stellar collapses is

$$N_{Gal} = (5.3 \pm 1.7) \ h^2 \qquad\qquad (1.6.2)$$

In the last millennium 6 supernovæ were optically observed in our Galaxy; that of the Lupus is a type Ia supernova and must be excluded from this evaluation. A lower limit of 1 galactic stellar collapse every 200 years can be set immediately. However, these *"Historical Supernovæ"* are clustered within a distance $D < 4 \ Kpc$ from the Sun; making some assumptions about the distribution of the stars in our Galaxy, more realistic limits can be set, but the uncertainties on this distribution cause again large uncertainties in the evaluation. Ratnatunga and Van den Bergh [RA89] obtained $N_{Gal} \approx 11 \div 14$, assuming that the census of the galactic stellar collapses is complete out to a radius of 3 Kpc, and $N_{Gal} \approx 6 \div 8$ assuming that it is complete out to a radius of 4 Kpc. Using a different model of the star distribution in the Milky Way, Van den Bergh lowered the second estimated rate to $N_{Gal} = 2.6 \pm 1.3$ [VA93]. On the other hand, taking into account the fact that the Historical Supernovæ are located within a viewing angle $i \approx 50°$, Tammann [TA82] obtained $N_{Gal} = 5.8 \pm 2.4$.

Since the residual of a stellar gravitational collapse is usually a neutron star and the galactic pulsars are rotating, highly magnetized neutron stars, the rate of the stellar collapses in the Galaxy can be estimated by measuring the rate of galactic pulsar formation. Using systematic surveys, Narayan [NA87] found a mean formation period P cf galactic pulsars

$$20 \ years \lesssim P \lesssim 60 \ years \tag{1.6.3}$$

(the preferred value is near the upper limit). Different studies gave different values of P (the lowest is $P \sim 6$); moreover, the fraction of neutron stars which are observable pulsars is an unknown parameter.

A direct calculation of the rate of galactic stellar collapses is possible (at least in principle) using the distribution (as a function of the mass) of the stars in the Galaxy $\Phi(M)$ and the rate $T(M)$ at which they are evolving. The frequency F of the galactic stellar collapses is given by the integral:

$$F(M_{min}) = N \int_{M_{min}}^{\infty} dM \ \frac{q\left[R(M)\right] f(M) \Phi(M)}{T(M)} \tag{1.6.4}$$

where N is the total number of stars in the Galaxy, $\Phi(M)$ and $T(M)$ have been defined above, M_{min} is the minimum mass of a star which explodes by core collapse, $q\left[R(M)\right]$ is the fraction of the stars of the Galaxy within a distance R from the Sun and $f(M)$ is the fraction of stars of mass M which collapse. The function q depends on the distance R only, each relation between R and M being absorbed in the star distribution $\Phi(M)$; the function $q(R)$ in the Bahcall-Soneira Galaxy model [BA80] is shown in fig. (1.11). Note that about $\approx 40\%$ of the stars of our Galaxy are within 8.5 Kpc from the Sun (distance of the Galactic Center, [KE86]) and $\approx 95\%$ within 20 Kpc.

Choosing $M_{min} \approx 10 \ M_\odot$ and assuming that a collapse anywhere in the Galaxy can be detected ($q = f = 1$), Bahcall and Piran [BA83] calculated a rate of galactic stellar collapses per century

$$N_{Gal} \approx 9 \tag{1.6.5}$$

This method is independent on the Hubble constant H_0 and does not require corrections for obscuration, but has a large uncertainty due to the form of the function $\Phi(M)$, which is poorly known for $M > 5 \ M_\odot$. Using a different function $\Phi(M)$ Ratnatunga and Van den Bergh [RA89] obtained $N_{Gal} = 1.0^{+1.5}_{-0.6}$ and $N_{Gal} = 2.2^{+2.7}_{-1.6}$, assuming $M_{min} = 8 \ M_\odot$ and $M_{min} = 5 \ M_\odot$ respectively.

The results of these calculations are summarized in table (1.5). In the last column the probability to observe at least one galactic stellar collapse in 10 years of running is reported. This probability is calculated from the Poisson formula

$$P(N \geq 1) = 1 - P(0) = 1 - \exp\left(-RT\right) \tag{1.6.6}$$

where R is the galactic stellar collapse rate, $T = 10 \ years$ is the running time and $P(0)$ is the probability to not detect any supernova.

As a general remark, the interstellar gas exhaustion rate looks in better agreement with the lower core collapse rates and very hard to reconcile with

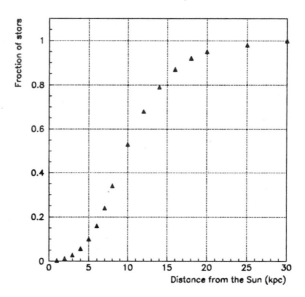

Figure 1.11: Bahcall-Soneira model of the fraction of the stars in our Galaxy as a function of the distance R from the Sun (after [BA80]).

the highest [VA91b] ($N_{Gal} \sim 10$). Therefore, an occurrence rate of a few supernovæ per century ($1/(20 \div 40)$ $years$) appears the most reasonable estimate.

Finally, the Baksan collaboration published an experimental limit on the galactic stellar collapse rate based on their galactic monitoring in the years 1980-1993 [BA94]. With a total 11.63 $year$ live time the limit

$$N_{Gal} < 19.8 \qquad (1.6.7)$$

at 90 % C. L. was set.

1.7   Detection of relic neutrinos from past supernovæ

Past supernovæ would have produced a diffuse neutrino background, if the supernova occurrence rate since the Big-Bang ($\approx 10^{10}$ $years$) has been larger or at least comparable to the present one; the neutrino flux and energy spectrum carry information on the evolution of the Universe.

Table 1.5: Number of galactic stellar collapses in a century calculated using different techniques; in the last column the corresponding probability to observe a galactic supernova in 10 *years* is reported.

Method	Number of collapses in a century	Probability of observation
Galactic and extragalactic supernova survey [VA91a]	$7.3\,h^2$	$17\,\%\,(h=0.5)$ $52\,\%\,(h=1)$
Galactic and extragalactic supernova survey [VA93]	$(5.3 \pm 1.7)\,h^2$	$30\,\% \div 50\,\%\,(h=1)$ $9\,\% \div 16\,\%\,(h=0.5)$
Historical Supernovæ [RA89]	$6 \div 8$ (4 Kpc) $11 \div 14$ (3 Kpc)	$45\,\% \div 55\,\%$ $67\,\% \div 75\,\%$
Historical Supernovæ [VA93]	2.6 ± 1.3 (4 Kpc)	$12\,\% \div 32\,\%$
Historical Supernovæ [TA82]	5.8 ± 2.4	$29\,\% \div 56\,\%$
Galactic pulsar formation rate [NA87]	$2 \div 5$	$18\,\% \div 39\,\%$
Massive star evolution [BA83]	9	$59\,\%$
Massive star evolution [RA89]	$1.0^{+1.5}_{-0.6}\,(M_{min} = 8\,M_\odot)$ $2.2^{+2.7}_{-1.6}\,(M_{min} = 5\,M_\odot)$	$4\,\% \div 22\,\%$ $6\,\% \div 40\,\%$

The expected relic neutrino background was calculated by many authors (e.g. [KR84], [WO86b], [TO95]); the results vary widely (by $2 \div 4$ orders of magnitude), though the arguments employed are rather similar.

The basic idea [KR84] is to consider the flux and energy spectrum of the neutrinos emitted by a standard supernova ($T_{\bar\nu_e} \sim 4$ MeV, $E_{tot} \approx 3 \times 10^{53}$ erg) and sum over all collapses occurred in the observable Universe. To do this, the expansion of the Universe must be taken into account: it produces a red-shift of the neutrino energy spectrum and a neutrino burst time broadening. A time integration must be performed from the Big-Bang till today; the result is a red-shift of a factor ranging from 1/2 (assuming a cosmological parameter $\Omega = 0$) to 3/5 (assuming $\Omega = 1$) on the mean neutrino energy. Krauss et al. [KR84], using a constant supernova occurrence rate of 1 stellar collapse every 15 years (a factor 2 overestimate of the favoured present supernova rate), obtained a $\bar\nu_e$ flux on the Earth

$$\phi(\bar\nu_e) \sim 25 \text{ cm}^{-2}\text{ s}^{-1} \tag{1.7.1}$$

with an average energy $\bar{E}_{\bar\nu_e} \sim 8$ MeV. On the other hand, Woosley et al. [WO86b] calculated the expected relic neutrino background following the des-

tiny of the stars in the mass range 10 $M_\odot \div 5 \times 10^5$ $M_\odot$ and using a supernova
current rate based upon galactic luminosity density estimates. For the present
supernova rate their evaluation agrees within a factor 3 with Krauss hypothesis, but the time integration of the red-shifted galactic luminosity widens the
difference, so that they obtained a conservative limit on the total neutrino flux
(i.e. summed over all neutrino flavours)

$$\phi \sim 1 \;\; \text{cm}^{-2} \; \text{s}^{-1} \tag{1.7.2}$$

Assuming the energy equipartition, the Woosley's $\bar{\nu}_e$ flux results about two
orders of magnitude lower than the Krauss $\bar{\nu}_e$ flux. Similar fluxes are expected
for the other flavour neutrinos.

The reason for considering only $\bar{\nu}_e$ is twofold: first of all, the present neutrino detectors are primarily sensitive to $\bar{\nu}_e$ (see next ch.), secondly the flux
(1.7.1) is 5 orders of magnitude lower than the ^8B solar ν_e flux, so that the
relic ν_e's are completely hidden by the more abundant solar neutrinos.

The spectrum of the relic neutrino background was systematically calculated in a recent paper [TO95] including a cosmological constant Λ; the result
was a further red-shift of the spectrum peak, which appeared at $\sim 3 \div 4$ MeV.
Using the calculated fluxes and the neutrino cross sections, the authors estimated the number of events which would be observed in one year in a water
Čerenkov detector in the range $10 \div 50$ MeV, where the background from the
other neutrino sources (reactor $\bar{\nu}_e$'s, atmospheric ν's, high-energy ν_e's from
the Sun) is an order of magnitude lower than the expected signal; the result is
~ 1 event/year/kton.

Experimental limits on the $\bar{\nu}_e$ flux were set by Kamiokande II and LS-
D collaborations, the first in the energy ranges $E > 9$ MeV [KA91b] and
19 MeV $< E_{\bar{\nu}_e} < 35$ MeV [KA88], the second in the energy range 12 MeV $<$
$E_{\bar{\nu}_e} < 30$ MeV [LS92]. The upper limits at $90\,\% \, C.L.$ are

$$\phi(\bar{\nu}_e) \leq 580 \;\; \text{cm}^{-2} \; \text{s}^{-1} \qquad\qquad E > 9 \;\; \text{MeV} \tag{1.7.3}$$
$$\phi(\bar{\nu}_e) \leq 212 \;\; \text{cm}^{-2} \; \text{s}^{-1} \qquad\quad 19 \;\; \text{MeV} < E < 35 \;\; \text{MeV} \tag{1.7.4}$$

These limits are well below the theoretical predictions but would be significantly improved by the future Super Kamiokande detector (≈ 50000 tonn H_2O,
see next ch.); following [TO95], this detector would observe ~ 25 events in one
year between 10 and 50 MeV in its 22000 tonn fiducial volume.

The upper limit on the $\nu_{\mu,\tau}$ flux ($\phi(\nu_{\mu,\tau}) \lesssim 3 \cdot 10^7$ cm^{-2} s^{-1} at $90\,\%$ C. L.
for 20 MeV $< E_{\nu_{\mu,\tau}} < 100$ MeV [LS92]) is much less stringent.

Chapter 2

State of the art in stellar collapse neutrino detection

A network of massive underground neutrino detectors is presently being assembled, whose combined sensitivity to a galactic neutrino burst will be unprecedented; it will be sensitive to one or more neutrino flavours and it will measure the total flux, the energy spectrum, the time structure, the flavour composition ... A coincident observation of a galactic stellar collapse by this detector network would give a very comprehensive picture of the stellar collapse mechanism and provide a great amount of information on neutrino physics.

Water Čerenkov's and scintillation counters are the most used experimental techniques for ν-detection, but some new generation experiments will also use Heavy Water, liquid Argon and higher mass number materials, as Na or Ca.

The expected neutrino signal in present and future detectors can be calculated from the models discussed in § (1.2) inserting in (1.4.4), (1.4.5) the appropriate neutrino cross sections and the characteristics of each experiment (total sensitive mass, type of active medium, efficiency ...).

2.1   Neutrino cross sections

The neutrino observation requires its conversion into an electrically charged particle or the excitation of nuclear levels, identified by the de-excitation radiation. The major difficulties in supernova neutrino detection are the smallness of the cross sections and the low energies involved ($E_\nu \lesssim 50$ MeV). In particular, the second point makes the neutrino elastic scattering on protons or nuclei a useless reaction, though this process has a relatively large cross section; a 10 MeV neutrino, incident on a target nucleus of mass number A, gives a nuclear recoil average kinetic energy of $T_{av} \approx 7 \times 10^{-2} A^{-1}$ MeV, practically undetectable.

The useful reactions are sensitive to one (charged current reactions) or to all (neutral current reactions) neutrino flavours; usually only a part of the

information on neutrino energy, direction and arrival time is preserved and
can be derived by a single process. Here we briefly review and discuss the ν
reactions used in supernova detectors.

a) The process

$$\bar{\nu}_e + p \rightarrow e^+ + n \qquad (2.1.1)$$

is the most important one in detectors based on Hydrogen rich media, like liquid scintillator and water. Its cross section is large and quadratically increases
with the energy E_{e^+} of the positron

$$\sigma_{\bar{\nu}_e p} \approx 8.5 \cdot 10^{-44} \, (E_{e^+}(\text{MeV}))^2 \, \text{cm}^2 \qquad (2.1.2)$$

(it is assumed: $m_n, m_p \gg E_{\bar{\nu}_e} \gg m_e$ where m_e, m_n and m_p are the electron, neutron and proton masses and $E_{\bar{\nu}_e}$ is the electron antineutrino energy)
[BA89a]. The e^+ comes to rest in the sensitive medium and annihilates with
an electron, giving a pair of γ's, with a total energy $E_\gamma = 2m_e = 1.022$ MeV;
this additional energy can also be detected.

The positrons emitted in (2.1.1) have a roughly isotropic angular distribu-
tion $(d\sigma/d\Omega_{\bar{\nu}_e,e^+} \approx 1 - 0.1 \cos(\theta_{\bar{\nu}_e,e^+})$ [BA89a]), so that no information on the
incident neutrino direction can be obtained from this process; the $\bar{\nu}_e$ energy
can be derived from the measured e^+ energy by the formula

$$E_{e^+} \approx E_{\bar{\nu}_e} - m_n + m_p - m_e \approx E_{\bar{\nu}_e} - 1.8 \text{ MeV} \qquad (2.1.3)$$

The neutron emitted in (2.1.1) is moderated by elastic scatterings on protons
and can be finally captured by a proton, forming a Deuterium nucleus. The
Deuterium binding energy is released as a 2.2 MeV photon (from now on: $\gamma_{2.2}$)

$$n + p \rightarrow d + \gamma_{2.2} \qquad (2.1.4)$$

The $\gamma_{2.2}$ can be detected by its secondary Compton electrons.

b) The neutrino elastic scattering on electrons

$$\nu_x, \bar{\nu}_x + e^- \rightarrow \nu_x, \bar{\nu}_x + e^- \qquad (2.1.5)$$

(where $x = e, \mu, \tau$) has a small cross section, which depends on the neutrino
flavour and linearly increases with the incident neutrino energy E_ν [BA89a]

$$\sigma_{\nu_x e} \approx \begin{cases} 9.2 \cdot 10^{-45} \, E_{\nu_e}(\text{MeV}) \, \text{cm}^2 & \text{if } x = e \\ 1.6 \cdot 10^{-45} \, E_{\nu_{\mu,\tau}}(\text{MeV}) \, \text{cm}^2 & \text{if } x = \mu, \tau \end{cases} \qquad (2.1.6)$$

$$\sigma_{\bar{\nu}_x e} \approx \begin{cases} 3.4 \cdot 10^{-45} \, E_{\bar{\nu}_e}(\text{MeV}) \, \text{cm}^2 & \text{if } x = e \\ 1.6 \cdot 10^{-45} \, E_{\bar{\nu}_{\mu,\tau}}(\text{MeV}) \, \text{cm}^2 & \text{if } x = \mu, \tau \end{cases} \qquad (2.1.7)$$

(the dependence on the ν flavour comes from the charged current diagram contribution, which is present for ν_e and $\bar{\nu}_e$, with different couplings, and absent

for $\nu_{\mu,\tau}$ and $\bar{\nu}_{\mu,\tau}$). The reactions (2.1.5) and (2.1.1) are completely different from a kinematical point of view: in (2.1.5) all the masses are negligible compared with the incident neutrino energy, while in (2.1.1) the baryon masses can be regarded as infinite compared with this energy. Therefore, in (2.1.5) a large fraction of the incident neutrino energy is carried away by the scattered neutrino.

The recoil electron angular distribution is strongly peaked in the incident neutrino direction (the electron emission angle is typically $\theta_e \sim (m_e/E_\nu)^{1/2}$), so that information on the supernova position can be obtained, if a sufficiently large number of (2.1.5) events are recorded and identified. The feasibility of this measurement was demonstrated by the Kamiokande II solar neutrino observation ([KA90], [KA91a]). The incident neutrino energy E_ν, the recoil electron energy E_e and scattering angle θ_e are related by the formula [BA89a]

$$E_\nu (\cos \theta_e) = \frac{E_e + |\cos \theta_e| \sqrt{E_e (E_e + 2)}}{[(E_e + 2)\cos^2 \theta_e - E_e]} \qquad (2.1.8)$$

where all the energies are given in units of m_e; therefore, the incident neutrino energy can be derived from the simultaneous measurements of E_e and θ_e.

The ratio between the number of (2.1.1) and ν_e-induced (2.1.5) events is sensitive to the neutrino energy spectra and is useful in discriminating between various supernova models [SC88].

c) The neutral current reactions of neutrinos and antineutrinos on nuclei include a long list of processes, whose cross sections are rather difficult to calculate because of nuclear structure effects.

In these reactions a neutron is knocked off or a nuclear level is excited; this level decays to the ground state by emission of radiation. The total energy of the de-excitation γ's is equal to the energy of the level, the energy of the neutrons is $\ll 1$ MeV and the recoil energy of the nuclei is negligible because of kinematical constraints. Then, the information of neutrino energy and arrival direction is lost, while the neutrino flux and arrival times can be measured.

The reactions on Carbon (present in liquid scintillators) were studied by many authors ([DO79], [MI82], [FU88]). The cross section for the flavour-blind superallowed (isovector, axial vector) transition

$$\nu_x, \bar{\nu}_x \ + \ ^{12}C(0^+,0) \ \rightarrow \ \nu_x, \bar{\nu}_x \ + \ ^{12}C^*(1^+,1)$$

$$\Downarrow$$

$$^{12}C(0^+,0) \ + \ \gamma(15.1 \ \text{MeV})$$

$$(2.1.9)$$

was recently measured by the KARMEN collaboration ([KA93], [KA94]) and the experimental value is in agreement with the [FU88] calculation. Because of its high energy threshold ($E_{th} = 15.1$ MeV), the process (2.1.9) is a good selector of $\nu_{\mu,\tau}$, $\bar{\nu}_{\mu,\tau}$, whose spectra are harder than that of ν_e, $\bar{\nu}_e$ (see figs. (1.4), (1.6)); therefore, (2.1.9) can give a direct measurement of the $\nu_{\mu,\tau}$, $\bar{\nu}_{\mu,\tau}$ flux. From now on we will denote the γ (15.1 MeV) by $\gamma_{15.1}$.

The excitation of ^{16}O has a quite small cross-section at the supernova neutrino energies [HA87] and can be neglected (it would give $\ll 1\%$ of the total number of events in a water based detector).

The most promising reaction for the neutral current detection is the neutrino induced Deuterium break-up ([BA88a], [HA89], [BA89b])

$$\nu_x, \bar{\nu}_x + d \to n + p + \nu_x, \bar{\nu}_x \quad (E_{th} = 2.2\,\text{MeV}) \qquad (2.1.10)$$

whose cross section is competitive with that of the (2.1.1) process. This reaction has a low energy threshold; then, the Deuterium disintegration can be induced also by the less energetic ν_e's, $\bar{\nu}_e$'s. The (2.1.10) process was suggested to measure the higher-energy ^8B ($E \leq 15$ MeV) and *hep* ($E \leq 18.5$ MeV) solar neutrino fluxes even if these neutrinos oscillate, in matter or in vacuum, before reaching the Earth.

Recently some authors ([HA88b], [CL90]) emphasized the good perspectives in the neutral current detection related to the use of heavy nuclei as target for neutrinos. The suggested reaction is

$$\nu_x, \bar{\nu}_x + (A, Z) \to \nu_x, \bar{\nu}_x + (A-1, Z) + \text{n} \qquad (2.1.11)$$

where "A" is a "heavy" nucleus. Calculations by Cline et al. [CL94] show that the cross section of the neutron knock-off process (2.1.11) is a steeply increasing function of the neutrino energy and it becomes larger than (2.1.1) at $E_\nu >$ 20 MeV, where a great enhancement due to the giant resonance excitation occurs; the heavy nuclei are therefore good selectors of the more energetic $\nu_{\mu,\tau}$, $\bar{\nu}_{\mu,\tau}$. The value of the (2.1.11) cross section, averaged on a supernova neutrino spectrum with a temperature $T = 8$ MeV, is $\sigma \sim 10^{-42}$ cm^2 per nucleon; these calculations have no experimental support.

d) The electron neutrinos and antineutrinos can interact with nuclei also via charged current reactions, converting into an electron or into a positron. Because of the mass difference between nuclei and electrons, the neutrino (antineutrino) and electron (positron) energies are related by a simple formula

$$E_{e^-,e^+} \simeq E_{\nu_e,\bar{\nu}_e} - Q \qquad (2.1.12)$$

where Q is the "Q-value" of the process, depending on the energy difference between the initial and final state (an example is (2.1.3)). From (2.1.12) the incident ν_e ($\bar{\nu}_e$) energy can be inferred. The electron (positron) angular distributions have the form

$$\frac{d\sigma}{d\Omega_{\nu e}} \propto 1 + \alpha\beta \cos\theta_{\nu e} \qquad (2.1.13)$$

where $\theta_{\nu e}$ is the angle formed by the incident neutrino (antineutrino) and electron (positron) directions, β is the electron (positron) velocity (≈ 1) and α

($|\alpha| \leq 1$) depends on the details of the interaction matrix element [BA89a] and (eventually) on the incident neutrino (antineutrino) energy. This distribution produces a forward - backward asymmetry. The charged current processes involving ν_e can be used for detecting and timing the infall-neutronization burst.

On Carbon (i.e. in liquid scintillator experiments) the following processes take place

$$\nu_e + {}^{12}C \rightarrow {}^{12}N + e^- \qquad (E_{th} = 17.3 \text{ MeV})$$

$$\Downarrow$$

$$^{12}N \rightarrow {}^{12}C + e^+ + \nu_e \qquad (\tau = 15.9 \text{ ms})$$

$$(2.1.14)$$

$$\bar{\nu}_e + {}^{12}C \rightarrow {}^{12}B + e^+ \qquad (E_{th} = 14.4 \text{ MeV})$$

$$\Downarrow$$

$$^{12}B \rightarrow {}^{12}C + e^- + \bar{\nu}_e \qquad (\tau = 29.3 \text{ ms})$$

$$(2.1.15)$$

where E_{th} and τ are the energy threshold of the reaction and the life time of the unstable nuclear states ^{12}N and ^{12}B. The cross sections for (2.1.14) and (2.1.15) processes are competitive with that of (2.1.9) at energies $E_\nu \gtrsim 30$ MeV; the measurements of the KARMEN collaboration are in agreement with the theoretical calculations for the (2.1.14) cross section [KA94].

The number of expected events for these reactions is low (only a few /kton of liquid scintillator), because of the high energy threshold E_{th}; only a small fraction of the "soft" ν_e, $\bar{\nu}_e$ spectrum can be associated with (2.1.14), (2.1.15) processes.

On the Oxygen nuclei the following charged current interactions are possible [HA87]

$$\nu_e + {}^{16}O \rightarrow {}^{16}F + e^- \quad (E_{th} = 15.4 \text{ MeV}) \qquad (2.1.16)$$

$$\nu_e + {}^{18}O \rightarrow {}^{18}F + e^- \quad (E_{th} = 1.66 \text{ MeV}) \qquad (2.1.17)$$

$$\bar{\nu}_e + {}^{16}O \rightarrow {}^{16}N + e^+ \quad (E_{th} = 11.4 \text{ MeV}) \qquad (2.1.18)$$

$$\bar{\nu}_e + {}^{18}O \rightarrow {}^{18}N + e^+ \quad (E_{th} = 14.1 \text{ MeV}) \qquad (2.1.19)$$

The number of events from the (2.1.16) and (2.1.18) reactions is limited by the high energy threshold; the (2.1.17) and (2.1.19) reactions are unimportant, because the relative abundance of ^{18}O is $\approx 0.2\%$ [LE78]. The cross sections for these processes are competitive with that of the (2.1.5) at energies $E \gtrsim 30$ MeV; this can have important consequences in case of neutrino mixing (see later). The angular distributions follow the (2.1.13), with α depending on the neutrino energy; for a Fermi-Dirac neutrino spectrum with $T = 5$ MeV, $\langle \alpha \rangle \approx -0.8$ [QI94]. The recoil electrons (positrons) are preferentially backward emitted.

The charged current reactions on Deuterium would give a substantial contribution to the total number of events in a heavy water experiment, because their cross sections are larger than those of the neutral current interaction (2.1.10) for $E_\nu > 10$ MeV ([BA88a], [HA89]). The possible charged current reactions on Deuterium are

$$\nu_e + d \rightarrow \text{p} + \text{p} + e^- \quad (E_{th} = 1.44 \text{ MeV}) \qquad (2.1.20)$$

$$\bar{\nu}_e + d \rightarrow \text{n} + \text{n} + e^+ \quad (E_{th} = 4.03 \text{ MeV}) \qquad (2.1.21)$$

Their energy thresholds are well below the mean energy of ν_e, $\bar{\nu}_e$ from a stellar collapse; their angular distributions are given by the (2.1.13) with $\alpha = -1/3$, which produces a $2 : 1$ backward - forward asymmetry.

Finally, the liquid Ar has a good sensitivity to ν_e via the charged current reaction

$$\left.\begin{array}{c} \nu_e + {}^{40}Ar \rightarrow {}^{40}K^* + e^- \quad (E_{th} = 5.885 \text{ MeV}) \\ \\ \Downarrow \\ \\ {}^{40}K + \gamma \quad (E_\gamma = 5 \text{ MeV}) \end{array}\right\} \qquad (2.1.22)$$

which has a cross section higher than (2.1.1) at $E_\nu > 20$ MeV [BA89a]. The recoil electrons are preferentially forward emitted ($\alpha = 1$ in (2.1.13)).

Fig. (2.1) shows the cross sections for the processes (2.1.1), (2.1.6) and (2.1.9) and fig. (2.2) for the processes (2.1.10), (2.1.20) and (2.1.21).

2.2   Second generation and near future supernova neutrino detectors

The supernova neutrino detectors have some common characteristics; the most obvious are the large sensitive mass (direct consequence of the small ν cross sections) and the low energy threshold ($\sim 5 \div 10$ MeV). The detectors must be located underground to reduce the cosmic ray background; a low radioactivity environment or some external shields against the natural background are needed to operate at the requested low threshold. The sensitive medium must be strong, as cheap as possible and have a low maintenance cost. The main quality factors are short dead-times, good relative timing and good energy resolution.

Today, water Čerenkov and liquid scintillator detectors are the only operating stellar collapse neutrino observatories, but in the "near future" the

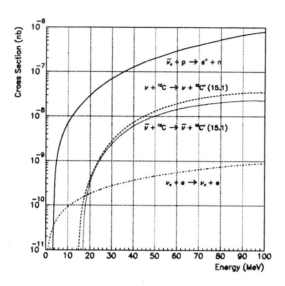

Figure 2.1: Cross sections for $\bar{\nu}_e$ charged current reactions on protons, ν_x and $\bar{\nu}_x$ neutral current reactions on Carbon nuclei and ν_e elastic scattering on electrons.

experiments will be more diversified and will be based on all reactions listed above. Here we discuss the relevant properties of the main experiments of each category.

2.2.1   Water Čerenkov detectors

Water Čerenkov detectors use large volumes of highly purified water, equipped with an array of inward-looking phototubes to detect the Čerenkov light produced by relativistic charged particles. The energy and direction of the particles can be inferred by the total amount of Čerenkov light and by the pattern of the illuminated PMT's. The Čerenkov detectors have a continuous active medium; they are self-shielded, i.e. the inner part of the detector is shielded from the external radioactivity background by the outer one; a fiducial volume can then be defined. The energy threshold of these experiments is in the range $5 \div 10$ MeV and their energy and angular resolution for $e^{\pm}$ are $\sigma(E)/E \sim 30 \div 60\,\%/\sqrt{E\,(\,\mathrm{MeV})}$ and $\sigma_\theta \sim 30°$ at $E = 10$ MeV. Water Čerenkov's are mostly sensitive to $\bar{\nu}_e$ via the (2.1.1) process; a few percent of

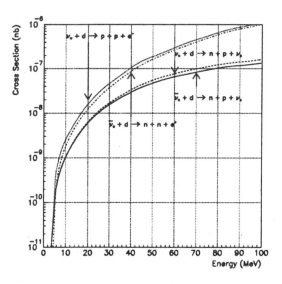

Figure 2.2: Cross sections for ν_x and $\bar{\nu}_x$ Deuterium disintegration and ν_e and $\bar{\nu}_e$ charged current reactions on Deuterium.

all events are due to neutrino-electron scatterings and to ν_e ($\bar{\nu}_e$) charged current reactions on Oxygen nuclei. Well-known examples of these experiments are IMB3 and Kamiokande II; a second generation water Čerenkov detector, *Super Kamiokande* ([TO87], [KO92]), is presently under construction in the Kamioka mine, the same site of Kamiokande II, at a 2700 *m.w.e.* (meters of water equivalent) depth.

Super Kamiokande will have a total sensitive mass M_{tot} = 50000 tonn of water, observed by 11200 ($\varnothing$ = 20 inch) PMT's. The fiducial mass for the stellar collapse detection will be 32000 tonn, the photocathode coverage 40 % of the total surface, the energy threshold $E_{th} \sim 5$ MeV, the angular resolution $\sigma_\theta \approx 27°$ and the energy resolution $\sigma(E)/E \approx 14\%$ at 10 MeV. Some thousand $\bar{\nu}_e$ events are expected in Super Kamiokande for a galactic stellar collapse; the angular resolution is adequate to distinguish a group of $\sim 100 \div 200$ electron scattering events, giving a 2° accuracy in "pointing-back" to the supernova. In principle, a further separation would be possible inside this group, because the $\nu_{\mu,\tau}$ induced events should cluster at a higher mean energy than the ν_e induced events [KO92]. The number of events in the lower and higher energy clusters should be comparable, making possible their statistical separation. The time distribution of the $\nu_e e$ scattering events should present a short initial peak due to the neutrinos emitted during the infall-neutronization phase. A layout of the Super Kamiokande experiment is shown in fig. (2.3). The expected number

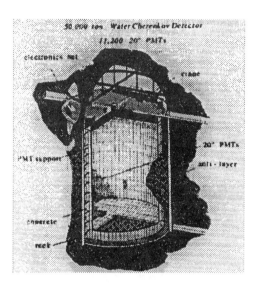

Figure 2.3: General layout of the Super Kamiokande experiment.

of events in Super Kamiokande for a collapse at 8.5 Kpc is reported in table (2.1); the model [BU92] is used. Note that the Oxygen reactions represent $\sim 1.5\,\%$ of the total number of events. They could be separated by the (2.1.1) only by looking at the angular distributions and searching for an excess in the backward direction. Supposing (unrealistically) that all the (2.1.16) electrons and (2.1.18) positrons are concentrated in the backward emisphere, this gives a ≈ 110 event excess in this emisphere. With a total number $\gtrsim 7500$ expected events, a $1\ \sigma$ statistical fluctuation would give a $\gtrsim 45$ event excess in each emisphere. The Oxygen signal statistical separation seems rather difficult.

2.2.2   Liquid scintillator detectors

Liquid scintillator detectors use large masses ($M \sim 10^3$ tonn) of clear hydrocarbons, segmented in some hundred counters (observed by two or more PMT's for each) or enclosed in a container and observed by an array of PMT's at the boundary of the active volume. Good timing ($\sigma_t \sim 1$ ns) and energy ($\sigma_E/E \sim 10\,\%$ at 10 MeV) resolutions are the most important features of these detectors. The segmented scintillator detectors have a good compatibility with tracking systems; this is important for some physical goals of these experiments (like the search for magnetic monopoles or the selection of up-

Table 2.1: Expected number of events in the Super Kamiokande experiment from a stellar collapse at the Galactic Center; the model [BU92] is used. ν_x indicates the sum of $\nu_{\mu,\tau}$ and $\bar{\nu}_{\mu,\tau}$.

Reaction	Events	Percentage of the total number of events
$\bar{\nu}_e + p$	7349	95.9
$\nu_e + e$	107	1.4
$\bar{\nu}_e + e$	23	0.3
$\nu_x + e$	69	0.9
$\nu_e + O$	50	0.65
$\bar{\nu}_e + O$	63	0.85
Total on electrons	199	2.6
Total on Oxygen	113	1.5
Total on protons	7349	95.9
Total	7661	100

going μ's), which are best studied by the coincident information coming from scintillation counters (dE/dx, timing) and from the tracking apparatus (trajectory). In the search for stellar collapse neutrinos the tracking information is helpful to identify and reject the cosmic ray background; the muons usually produce time and space correlated signals in the scintillation counters and in the tracking elements.

The scintillation counters have generally a larger light yield and therefore a higher sensitivity than the water Čerenkov's; this makes the $\gamma_{2.2}$ (2.1.4) detectable in scintillation experiments. The neutron moderation time in liquid scintillator is $\approx 10\ \mu s$, the neutron capture time $\approx 180\ \mu s$. The $\gamma_{2.2}$ gives a powerful further signature of the e^+ event for the reaction (2.1.1).

Fig. (2.4) shows the *LVD* [LV88] (**Large Volume Detector**) experiment; this detector was designed to have a 1800 tonn liquid scintillator total mass; its present active mass is ≈ 600 tonn (spring 1995).

LVD is located in the Hall "A" of the Gran Sasso National Laboratory (see next ch.); it is a modular detector; it will be made of 190 modules arranged in 5 towers; each module contains 8 scintillation counters, $1 \times 1.5 \times 1\ m^3$ in size. The scintillator volume is interleaved by a streamer tube tracking system. Each counter is observed by three PMT's ($\emptyset = 15$ cm); the liquid scintillator density is 0.8 g/cm^3, its attenuation length > 15 m. The trigger is given by a three fold coincidence of the PMT signals. A double energy threshold can be set: a higher primary (~ 7 MeV) to detect positron-like events and a lower secondary (~ 1 MeV) to detect the $\gamma_{2.2}$'s. The energy resolution and the

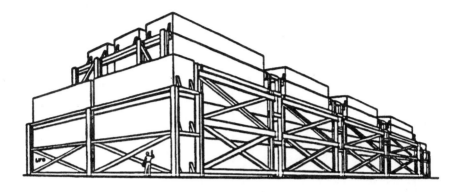

Figure 2.4: The LVD detector in its final configuration (after [LV88]).

relative timing accuracy are $\sigma(E)/E = 15\%$ at $E = 10$ MeV and 12.5 ns.

LVD has a partially self-shielded structure: the counters can be divided in external (216 per tower) and internal (88 per tower); there are also external shields of borax paraffin (a 10 cm layer) and iron (a 2 cm layer) between the counters. The average background rate is $\sim 3 \div 4 \times 10^{-4}$ Hz per counter for an energy release $E > 7$ MeV and ~ 200 Hz per counter for $E > 1.5$ MeV; if only the internal counters are considered, the average background rate at the secondary thresholds is ~ 40 Hz per counter, a factor 5 improvement due to the shielding. The detection efficiency for the $\gamma_{2.2}$ is $\approx 70\%$; the expected signal to noise ratio for the secondary signals is ≈ 5 in a 600 μs time window after the primary trigger ([LV92], [LV93]).

Table (2.2) lists the number of expected events for the reactions (2.1.1), (2.1.14), (2.1.15), (2.1.5) and (2.1.9) from a supernova at the Galactic Center in MACRO (600 *tonn* of liquid scintillator) with a 7 MeV energy threshold, calculated using the models discussed in § (1.2) (for [JA93] paper we adopted the central values for $\langle E \rangle$, η and f and a total energy released in neutrinos $E_{tot} \approx 2.7 \times 10^{53}$ ergs) and the formula (1.4.4). Apart from the [NA80] optimistic predictions, the expected number of (2.1.1) events in MACRO for a standard stellar collapse at the Galactic Center is ≈ 200; the fraction of $\bar{\nu}_e\,p$ events is $> 92\%$ for all the models; C events contribute $\sim 6\%$ and the electron scattering makes for the remaining part. Practically all the events are expected from the cooling stage. An extension of these calculations to different mass detectors (LVD in its final configuration, LSD, Baksan ...) is straightforward,

Table 2.2: Number of events from a stellar collapse at the Galactic Center expected in MACRO. For electron scattering and neutral current reactions ν_x is equivalent to the sum of all ν and $\bar{\nu}$ flavours. Some different models ([BL88], [NA80], [BU92], [BR91], [JA93]) were used.

Model	Reaction				
	$\bar{\nu}_e p$	$\nu_x, \bar{\nu}_x {}^{12}C$	$\nu_x e$	$\nu_e {}^{12}C$	$\bar{\nu}_e {}^{12}C$
Nadezhin & Otroshchenko [NA80]	453	27	6	$3 \div 4$	$4 \div 5$
Bludmann & Schindler (constant T model) [BL88]	182	9	2	1	1
Bludmann & Schindler (ν-sphere cooling) [BL88]	247	12	3	1	1
Bruenn & Haxton [BR91]	209	11	4	1	1
Burrows, Klein & Gandhi [BU92]	199	4	3	< 1	< 1
Janka & Hillebrandt [JA93]	206	9	4	< 1	< 1

except for (2.1.9), for which the detection efficiency largely varies from experiment to experiment: for instance, for the LVD counters it would be $\sim 50\,\%$, while for MACRO counters is $\sim 30\,\%$ [BA90a]. This efficiency was taken into account in table (2.2). The charged current events on ^{12}C give $< 1\,\%$ of the total number of events, but have a good signature because of the subsequent β-decays with high energy electrons and positrons (see (2.1.14) and (2.1.15)). The efficiency for detecting these events with a 7 MeV energy threshold would be $\sim 50 \div 60\,\%$.

2.2.3   Heavy water Čerenkov detectors

SNO (Sudbury Neutrino Observatory [SN88], fig. (2.5)), a large heavy water detector, will consist of 1000 tonn of D_2O, surrounded by a 5000 tonn light water shield (the inner fiducial mass of 1.6 kton H_2O can also be used for supernova neutrino detection). It will be located in the Creighton mine (Canada), at a 5900 $m.w.e.$ depth. The heavy water will be observed by 9500 PMT's; the energy and angular resolution would be $\sigma(E)/E \approx 16\,\%$ and $\sigma_\theta \approx 30°$ at 7 MeV, the energy threshold ≈ 5 MeV. SNO will be sensitive to the neutral current interactions (2.1.10) on Deuterium; the secondary neutrons will be detected by ^3He counters (which use the process $^3He + n \rightarrow t + p + 0.764$ MeV) or will be captured by ^{35}Cl (salted in form of $MgCl_2$ or $NaCl$). The neutron capture leaves the ^{36}Cl in an excited state; the subsequent de-excitation will release a

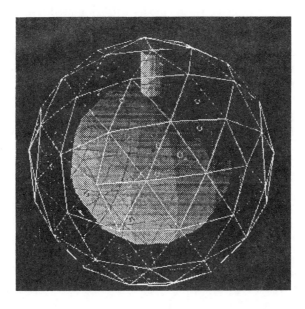

Figure 2.5: The SNO detector

γ-ray cascade with a total energy $\sum_i E_i \approx 8.6$ MeV, which will be detected through the Čerenkov light produced by the secondary Compton electrons. The efficiency for this reaction was recently measured [SN94] using a ^{252}Cf source; with a 0.25 % (in weight) $NaCl$ doping, $\epsilon \approx 60\%$ was obtained. The neutron capture time with this doping is $\tau \approx 5$ ms.

The γ background is a serious problem in SNO, since the neutral current event (2.1.10) can be simulated by a Deuterium nucleus disintegration induced by a photon having an energy $E_\gamma > 2.2$ MeV (this disintegration process is the inverse of (2.1.4)). The heavy and light water are purified at a level of 10^{-14} g/g and the building materials must be accurately selected on the basis of their radioactive contamination.

The flavour blind process (2.1.10) will measure the total neutrino flux, while the ν_e, $\bar{\nu}_e$ total flux and spectra will be measured via the charged current reactions (2.1.20) and (2.1.21) in D_2O and (2.1.1) in the H_2O fiducial volume. The separation between electrons from (2.1.20) and positrons from (2.1.21) is possible by using the positron-two neutron signal coincidence in (2.1.21) and the neutron absence in (2.1.20); this technique can be compromised by the abundant neutrons (2.1.10). All these reactions will also give the event time information; for the neutral current reaction, the accuracy of the time measurement will be limited because of the long neutron capture time.

The number of expected events in SNO from a stellar collapse at the Galactic Center is given in table (2.3); the model [BU92] is used. Note that in the

Table 2.3: Expected number of events in the SNO experiment from a stellar collapse at the Galactic Center; the model [BU92] is used. ν_x indicates the sum of $\nu_{\mu,\tau}$ and $\bar{\nu}_{\mu,\tau}$; the reactions on electrons and on Oxygen nuclei take place in H_2O and D_2O.

Reaction	Events	Percentage of the total number of events
$\bar{\nu}_e + p$	458	42.4
$\nu_e + e$	17	1.5
$\bar{\nu}_e + e$	4	0.37
$\nu_x + e$	9	0.83
$\nu_e + O$	$4 \div 5$	$0.37 \div 0.46$
$\bar{\nu}_e + O$	$5 \div 6$	$0.46 \div 0.55$
$\nu_e + d\,(CC)$	113	10.4
$\bar{\nu}_e + d\,(CC)$	92	8.5
$\nu_e + d\,(NC)$	49	4.5
$\bar{\nu}_e + d\,(NC)$	52	4.8
$\nu_\mu + d\,(NC)$	277	25.6
Total NC on Deuterium	378	35.0
Total CC on Deuterium	205	19.0
Total CC on Oxygen	10	0.9
Total on electrons	30	2.7
Total on protons	458	42.4
Total on D_2O	597	55.2
Total on H_2O	484	44.8
Total	1081	100

[BU92] paper a 100 % efficiency above the energy threshold is assumed; this assumption is acceptable for charged current reactions (the authors quote an overestimated error $\lesssim 15\,\%$), but uncorrect for the neutral current reaction. When the measured efficiency is included, the number of neutral current reactions reduces to ≈ 230.

About 18 ν_e events (16 on Deuterium and 2 on electrons) would be detected during the first 10 ms in the infall-neutronization burst phase.

The backward - forward asymmetry of the (2.1.20), (2.1.21) processes would produce a ≈ 18 event excess in the backward emisphere and a ≈ 18 event deficit in the forward emisphere. The total number of expected events in D_2O in each emisphere is ≈ 300 (220 including the neutron detection efficiency);

the corresponding 1 σ fluctuation is $\approx$ 9 ($\approx$ 8). The use of the (2.1.20), (2.1.21) reactions to point back to the supernova is limited by the statistical fluctuations.

2.2.4   Other detectors and future projects

ICARUS

The good sensitivity to ν_e electrons would be the major characteristic of *ICARUS* (Imaging of Cosmic And Rare Underground Signal [IC94]), a 5 kton liquid Argon time projection chamber (in project). In case of a stellar collapse at the Galactic Center ICARUS would detect 105 events (assuming a 11 MeV detection threshold for the (2.1.22) process [IC94]): 61 from (2.1.22), 25 from $\nu_e e$ scattering and 19 from $\bar{\nu}_e e$ and $\nu_x e$ scattering (ν_x indicates the sum of $\nu_{\mu,\tau}$ and $\bar{\nu}_{\mu,\tau}$). Therefore $\approx$ 82 % of the total number of events will be due to ν_e and 15 of them will be concentrated in the first 10 ms, giving a clean signature of the infall-neutronization burst. In principle, ν_x scattering events recorded in ICARUS could help setting limits on ν_x masses in the range m_{ν_x} $\gtrsim$ 150 eV; however, the expected number of ν_x induced events in ICARUS is rather small. The high-resolution tracking capabilities of this detector (spatial resolution $\sim$ 100 μm) would allow to point back to the supernova also with a limited ($\sim$ 100 events) statistics.

Ideas for detectors sensitive to extragalactic supernovæ

A simple inspection of tables (2.1), (2.2) and (2.3) shows that the second generation detectors are adequate only for galactic stellar collapse; for instance, a collapse at 1 Mpc (the distance of some Local Group Galaxies, *Andromeda* and *M33*) would produce less than 1 $\bar{\nu}_e$ event in Super Kamiokande. Therefore a detector for ν's from extragalactic supernovæ should have a mass $\gtrsim 10^6$ tonn H_2O, 20 times bigger than Super Kamiokande, to be sensitive to stellar collapses in the Local Group and $\gtrsim 10^8$ tonn H_2O (2000 times bigger than Super Kamiokande !) to be sensitive to stellar collapses in the Virgo Cluster (distance $\approx$ 10 Mpc). However, the rate of stellar collapses in the Local Group is $\sim$ 10 times higher than that in the Milky Way; then, an experiment sensitive to collapses in the Local Group would have a much higher probability to detect a supernova than an experiment restricted to our Galaxy. An other motivation for an extragalactic supernova ν detector is its sensitivity to small values of ν mass: a 5 eV neutrino mass would cause a $\delta t \approx$ 10 s arrival time difference between a 10 MeV and a 50 MeV neutrinos emitted simultaneously in a collapse at 1 Mpc distance (see (1.5.1)). Having this in mind, some ideas were proposed; we briefly discuss *SNBO* (Supernova Neutrino Burst Observatory, [CL93], [CL94]) and *DOUGHNUTS* (Detector Of Under-Ground Hideous Neutrinos from Universe and from Terrestrial Sources, [KO92]).

SNBO is an embryonic project of a detector based on a large A target material, suggested for taking advantage of the large neutron knock-off (2.1.11) cross sections.

SNBO would use $M \gtrsim 10^7$ tonn $NaCl$ as sensitive medium, it would be located in a salt mine and it would be instrumented with BF_3 neutron counters, interspaced in the $NaCl$ in a long ~ 1 Km tunnel. A conceptual design of this detector is shown in fig. (2.6). The estimated total number of events in SNBO

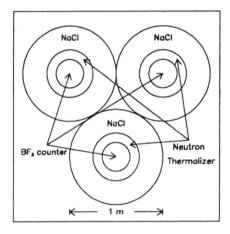

Figure 2.6: Conceptual design of the SNBO detector (after [CL90]).

is ~ 20 for a supernova at 1 Mpc distance; a precise calculation is impossible because of the large uncertainties on the cross sections (2.1.11) and on the neutron detection efficiency, which strongly depends on the geometrical design of the experiment. Such a detector requires a large natural deposit of $NaCl$ in a neutron low-background environment; some sites were proposed and are under testing. The applicability of the BF_3 counter technology on such a large scale must also be demonstrated.

DOUGHNUTS (fig. (2.7)) is a project for a large scale (1 Mton) water Čerenkov detector, consisting of three doughnut-shaped subdetectors, which eventually could be built in three steps. The total fiducial mass would be ≈ 800 kton, the total surface $\approx 1.5 \times 10^5$ m^2 and the total number of ($\emptyset = 20$ inch) PMT's ≈ 50000. With this PMT number, a $\approx 5\%$ photocathode coverage is achieved; the energy threshold would be ≈ 10 MeV. The photocatode coverage is a compromise between the cost and the performances (energy res-

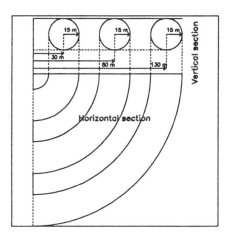

Figure 2.7: Conceptual design of the DOUGHNUTS detector (after [KO92]).

olution and e/μ discrimination); the estimated cost is ≈ 100 M\$. In case of a collapse at 1 Mpc distance DOUGHNUTS would record ≈ 15 events above the energy threshold and in case of a collapse at the Galactic Center $\gtrsim 150000$ events. This experiment could also search for proton decay and high-energy neutrino sources, detect solar ^8B neutrinos, explore the mass composition of the cosmic rays at energies $\gtrsim 10^{15}$ eV and study $\nu_\mu \leftrightarrow \nu_e$ and $\nu_\mu \leftrightarrow \nu_\tau$ oscillations by using an accelerator beam (τ appearance; the τ leptons could be identified looking for μ's from τ decay with a large (> 50 %) missing energy in respect to the neutrino beam direction).

High-Energy Neutrino Telescopes and supernovæ

Recently Halzen et al. [HA94] stressed that also the high energy neutrino telescopes (*HENT's*) can detect supernovæ. HENT's consist of arrays of ~ 200 Optical Modules (*OM*), placed in deep clear water (*DUMAND*, **D**eep **U**ndersea **M**uon **A**nd **N**eutrino **D**etector, *NESTOR*, **NE**utrinos from **S**upernovæ and **T**eV sources **O**cean **R**ange, *Baikal*) or ice (*AMANDA*, **A**ntaric **M**uon **A**nd **N**eutrino **D**etector **A**rray, fig. (2.8)), used as sensitive media and cosmic ray shield. The Čerenkov light emitted by energetic μ's or in electromagnetic showers is collected by a group of OM's, giving a coincident signal. Though the energy threshold of these detectors is in the GeV range, HENT's could detect supernovæ by using a collective effect: a stream of some thousand low

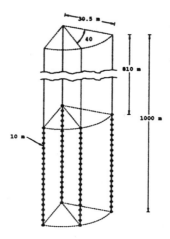

Figure 2.8: Layout of the AMANDA experiment.

energy positrons would suddenly yield signals in all OM's. A collapse at the Galactic Center would produce in AMANDA about 5000 events in 200 OM's; such a signal would have a $5\,\sigma$-statistical significance above the experimental background (~ 1 kHz/OM) and it would allow an independent confirmation of a supernova signal seen by other detectors.

2.3 Effects of non standard physics on the expected neutrino signal

If all neutrinos are massless, the luminosity profile and the energy spectrum of the neutrino events detected in the various experiments will follow that of the neutrino emission, folded with the individual response of each detector. If, on the other hand, at least one of the three neutrinos is massive, the signals in the experiments will be distorted by two effects:

- the flight time difference of neutrinos of different energies (see § (1.5.1) for the *SN1987A* case);

- the neutrino flavour mixing (in the supernova matter or in vacuum).

Other non standard physics predictions (ν-decay, ν magnetic moment etc.) would also affect the recorded neutrino signal; however, these predictions appear rather exotic in the light of the *SN1987A* and of the recent solar neutrino data ([GA92b], [GA94a]).

Effects of a finite neutrino mass

A finite $\bar{\nu}_e$ mass would smear, broaden and delay the dominant (2.1.1) signal in water Čerenkov and liquid scintillation detectors. The flight time difference is proportional to the supernova distance; so, for a galactic collapse the signal dispersion should be small; for instance, a 5 eV ν_e mass will cause a $\delta t \approx 0.1$ s separation between two neutrinos of 10 MeV and 50 MeV energies for a collapse at the Galactic Center. However, a high statistics $\bar{\nu}_e$ detection would compensate for the "short baseline" and allow to reach a sensitivity comparable to the $\bar{\nu}_e$ mass terrestrial limit ($m_{\bar{\nu}_e} \lesssim 4.8$ eV, [BE94]).

If we remind the ν luminosity spectrum (fig. (1.5)), we observe that the ν_e, $\bar{\nu}_e$ luminosities are characterized by modulations with a time-scale of ≈ 50 ms; such modulations would be manifest in an oscillating ν signal in case of massless $\bar{\nu}_e$ and would be masked or largely obscured by a $\bar{\nu}_e$ mass of few eV. Also the ν_e neutronization burst in SNO and ICARUS would be destroyed and the initial rise of the ν_e, $\bar{\nu}_e$ luminosity would be less sharp. Figure (2.9) shows the effects of a finite $\bar{\nu}_e$ mass on the signal in a liquid scintillation detector ("LVD") for a galactic stellar collapse [BU92]. Very similar effects are expected in a water Čerenkov detector. In principle an analysis for various masses could allow to unfold from the observed distribution the finite mass effect and extract the "best" m_{ν_e} value.

While the expected sensitivity of stellar collapse neutrino detectors to a ν_e mass is comparable with the limits obtained by the classical tritium β-decay experiments, a great improvement is expected for the $\nu_{\mu,\tau}$ mass limits. The terrestrial limits are $m_{\nu_\mu} \leq 160$ KeV [AS94] and $m_{\nu_\tau} \leq 24$ MeV [AL95], while crude order-of-magnitude considerations suggest a stellar collapse experiment sensitivity to $m_{\nu_{\mu,\tau}} \gtrsim 10 \div 20$ eV, a $\nu_{\mu,\tau}$ mass region extremely significant for cosmology.

Now suppose that ν_e is massless or at least $m_{\nu_e} \ll m_{\nu_\mu,\nu_\tau}$; neglecting for the moment the possibility of a neutrino mixing, the time of flight broadening effect would have a clear signature in SNO. The time evolution of the fraction of the deuteron breakup rate to the total number of events would be enhanced at "long" times ($t \gtrsim 1$ s) after the signal start [BU92] by a finite $\nu_{\mu,\tau}$ mass in the $0 \div 100$ eV range. This effect is shown in fig. (2.10). The time distribution of the neutral current (2.1.10) events would be broadened by a finite $\nu_{\mu,\tau}$ mass, but the effect would be partially masked, because a fraction of the (2.1.10) events is induced by the "massless" ν_e, $\bar{\nu}_e$. The sensitivity of the neutral current event time distribution to the $\nu_{\mu,\tau}$ mass is improved if the ν_e, $\bar{\nu}_e$ contributions could be statistically separated by using independent measurements of their

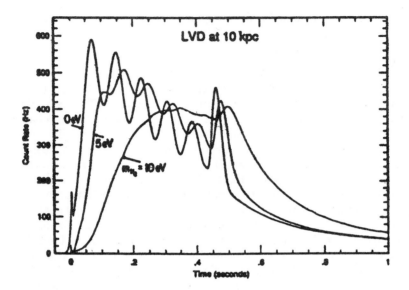

Figure 2.9: Expected signal in a liquid scintillation detector ("LVD") for a galactic stellar collapse calculated by Burrows et al. [BU92] for a $\bar{\nu}_e$ mass of 0, 5 and 10 eV.

fluxes and time structures based on the (2.1.20), (2.1.21) reactions.

Though liquid scintillation detectors are not particularly sensitive to the neutral current reactions, limits on ν_μ, ν_τ masses could be set by identifying a group of (2.1.9) events and looking at their time spread [RY92]. However, this identification is rather difficult, because the $\gamma_{15.1}$ event gives a signature equivalent to that of a $12 \div 18$ MeV e^+ event, unless the secondary $\gamma_{2.2}$ (2.1.4) is detected. The possibility of a statically significant separation of the (2.1.9) signal from the dominant one (2.1.1) sets the ν_μ, ν_τ mass lower sensitivity of such experiments; for a collapse at the Galactic Center the minimum observable mass in MACRO is ≈ 300 eV. The sensitivity upper limit is fixed by the experimental background, because for a too large ν_μ, ν_τ mass, the signal would be so much spread out as to be of difficult identification. This upper limit for MACRO is ≈ 1 KeV. The same limits for LVD (final configuration) would be $m_\nu^{min} \approx 100$ eV and $m_\nu^{max} \approx 4$ KeV [RY93]. A detailed discussion of the potentiality of this technique is given in [AC90].

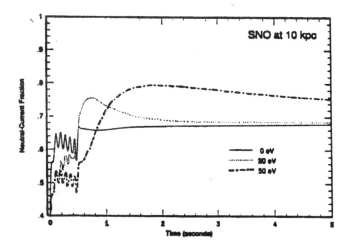

Figure 2.10: Effects of a massive neutrino in the $0 \div 50$ eV range on the neutral current event fraction in SNO (after [BU92]).

Effects of neutrino mixing

The supernova experiments are the longest baseline experiments for studying neutrino vacuum oscillations; the sensitivity in δm^2 reaches 10^{-19} eV^2, but the sensitivity in $\sin^2 (2\,\theta)$ is much poorer, because of the large uncertainties in the neutrino signal predictions.

A $\bar{\nu}_e \leftrightarrow \bar{\nu}_x$ vacuum mixing ($x = \mu, \tau$) would be observable, since the higher energy component of the $\bar{\nu}_e$ spectrum would be enriched [FU92]; therefore, the number of $\bar{\nu}_e$ events in liquid scintillator and water Čerenkov detectors would be larger than that expected in the case of no mixing and the $\bar{\nu}_e$ excess due to the $\bar{\nu}_x \to \bar{\nu}_e$ conversion would have a harder spectrum than the one of "unconverted" $\bar{\nu}_e$'s.

A $\nu_x \to \nu_e$ oscillation would produce a similar hardening effect on the ν_e signal in ν_e-sensitive experiments. In Super Kamiokande a signature of a $\nu_x \leftrightarrow \nu_e$ mixing would be given by an increase in the relative yield of the backward-peaked electrons (2.1.16) [QI94]; the statistical separation of these electrons from the more abundant (2.1.1) events by looking at their different angular distribution would be feasible for a strong $\nu_x \to \nu_e$ conversion.

A possible signature of a neutrino mixing would be given in liquid scintillator detectors by an increase of the (2.1.14), (2.1.15) events; in case of a total $\nu_x \to \nu_e$ and $\bar{\nu}_x \to \bar{\nu}_e$ conversion the number of these reactions would increase by a factor $\gtrsim 10$ [RY94]. The sensitivity limit of the neutrino mixing signature is set by the statistical fluctuations of the number of (2.1.14), (2.1.15) events induced by unconverted ν_e's and $\bar{\nu}_e$'s; for a collapse at the Galactic Center and a 1 kton liquid scintillation detector, the minimum number of charged current interactions on ^{12}C to be detected for a statistically significant (90 % C. L.) signature of neutrino oscillations is ≈ 18 ([RY94], model [NA80] used). The corresponding minimum mixing angle is $\sin^2(2\theta_{min}) \approx 0.28$.

The matter oscillations would cause somewhat different effects. The most important point is that the electron antineutrinos do not undergo matter oscillations, unless $m_{\nu_e} > m_{\nu_x}$, an unlikely possibility. Therefore, the $\bar{\nu}_e$ sensitive experiments (water Čerenkov's and liquid scintillators) could measure the unconverted $\bar{\nu}_e$ flux and spectrum, while the $\nu_x \rightarrow \nu_e$ conversion would be indicated by (2.1.16) events, as in case of vacuum oscillations. Moreover, the matter oscillations depend on the neutrino energy more strongly than the vacuum oscillations: in the former case, the resonant conversion in a supernova core occurs (at fixed δm^2) only in a narrow range of neutrino energies, in the latter the functional dependence on the neutrino energy E_ν ($P \propto \sin^2(\delta m^2 L/E_\nu)$) is completely smeared out and the conversion probability is a function of the mixing angle only. Theoretical calculations ([BU93], [QI94]) show that the MSW mechanism enhances the number of charged current (2.1.16) events in Super Kamiokande by a factor $\sim 4 \div 5$, making reliable their identification by means of the backward-peaked angular distribution; in SNO an increase by a factor ~ 1.5 is expected for the (2.1.20) electrons, together with a hardening of their energy spectrum. A large $\nu_e \leftrightarrow \nu_x$ conversion will also destroy the ν_e neutronization burst.

In liquid scintillator experiments also a matter oscillation could be signaled by a larger number of (2.1.14) reactions; the sensitivity region in the $(\delta m^2, \sin^2(2\theta))$ plane can be determined by requiring a 90 % C. L. statistical significance in the neutrino oscillation signature; see [RY94] for the calculation details.

In SNO the neutrino oscillations would produce an enrichment of the short-time component of the neutral current event temporal distribution, due to $\nu_x \rightarrow \nu_e$ conversion; this would make more difficult the $\nu_{\mu,\tau}$ mass measurement by the time of flight technique. A statistical subtraction of the ν_e contribution (to be determined by the (2.1.20) reaction) would improve the quality of this measurement; the subtraction requires an accurate knowledge of the experimental efficiencies for both the charged and neutral current reactions.

A conversion between ν_x ($x = e, \mu, \tau$) and a sterile neutrino ν_s would be signaled by a number of observed events lower than the expected in all detectors. The sensitivity to the $\nu_x \leftrightarrow \nu_s$ mixing is limited by the theoretical uncertainties on the expected signal.

Finally, it must be stressed that a $\nu_\mu \leftrightarrow \nu_\tau$ oscillation in matter or vacuum is undetectable in a supernova neutrino experiment, because ν_μ and ν_τ are experimentally indistinguishable (the neutrino energy is below the μ production threshold).

Chapter 3

The MACRO Detector: characteristics and performances

3.1 The Gran Sasso National Laboratory

The MACRO detector is located in the Hall "B" of the Gran Sasso National Laboratory [BE88] near L' Aquila (Italy), at a longitude of 13° 34' 28" East, a latitude of 42° 27' 09" North and an average height $\langle h \rangle \approx 960$ m on the sea level.

The laboratory is at the center of the Roma-Teramo highway tunnel; it has a rock average shield $\langle s_{av} \rangle \approx 3700$ $m.w.e.$; the minimum shield is $s_{min} \approx 3150$ $m.w.e.$. The rock above the tunnel attenuates the external cosmic ray flux by a factor $\approx 10^6$; the resulting μ flux is $\phi \approx 1\, m^{-2}\, h^{-1}$ [MA90] and the minimum muon energy requested to reach the Hall "B" is $E_{min} \approx 1.4$ TeV.

The laboratory contains three large area experimental halls (typical dimensions $100 \times 30 \times 20$ m^3), which can be easily entered by using the highway tunnel. The large reduction of the cosmic ray flux, the rock low radioactivity and its unique internal facilities make the Gran Sasso laboratory one of the most interesting site for low-background underground experiments. A laboratory map is shown in fig. (3.1).

3.2 MACRO physical goals

Most large size underground experiments are multi-purpose and MACRO is not an exception; it was built having several physical goals in mind [MA84], in addition to the search for stellar gravitational collapses.

The detector was primarily developed to search for supermassive ($M_{mon} \sim 10^{16}$ GeV/c^2) GUT magnetic monopoles ([HO74], [PO74], [PR84]) on a wide velocity range ($4 \times 10^{-5} < \beta_{mon} < 1$) at a sensitivity level one order of magni-

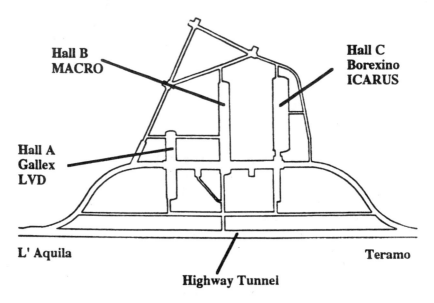

Figure 3.1: Map of the Gran Sasso laboratory.

tude below the Parker astrophysical bound [PA82]. A first limit on the slowly moving monopoles ($\beta < 4 \times 10^{-3}$) was already published [MA94a]; in fig. (3.2) this limit is presented and compared with the previous best limits ([BE90a], [HU90], [OR91], [BA82], [SO92], [BU90a]) and with the Parker bound. The limit which will be set by MACRO with a null monopole signal in 5 years of running is also shown.

The high-energy primary cosmic radiation ($E_{pr} \gtrsim 3$ TeV) can be studied indirectly in MACRO looking at the properties of the secondary muon showers, produced by the interactions of the primary radiation with the nuclei of the atmosphere. Some information on the primary composition can be obtained studying the surviving muon multiplicity distribution and comparing it with theoretical predictions [MA92c]; independent checks of the algorithms and of the hadronic cross sections used in these calculations can be performed looking at the distribution of the distance between muons in a pair (*"decoherence function"*) [MA92d]. Further information can be obtained from the correlated analysis ([ME90], [ME94]) of the MACRO events with those of the Extensive Air Shower detector *EAS-TOP* [EA89], located on the Gran Sasso surface (≈ 2200 m on the sea level).

The muon and neutrino astronomy is also one of the MACRO primary goals. An all-sky survey can be performed, searching for a muon excess in the direction of known high energy ($E_\gamma \gtrsim 2$ TeV) γ ray sources (like **Cyg-X3**); no signal was detected yet [MA93b]. The upper limits obtained were $J_\mu \lesssim 2 \times 10^{-12}$ cm^{-2} s^{-1} for a steady muon flux and $J_\mu \lesssim 8.8 \times 10^{-13}$ cm^{-2} s^{-1}

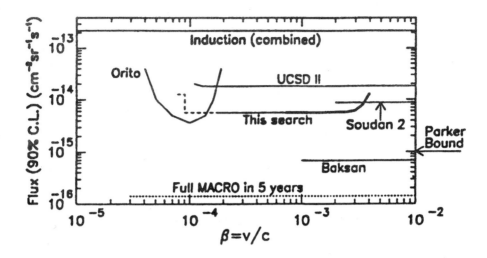

Figure 3.2: Experimental limits on the flux of magnetic monopoles ([BE90a], [HU90], [OR91], [BA82], [SO92], [BU90a]) compared with the Parker bound. The curve labeled with "This Search" represents the best MACRO limit on the slow monopole flux (adapted after [MA94a]).

for a modulated muon flux from **Cyg-X3** (the orbital motion of a source could introduce a modulation in the muon flux).

Upgoing muons (i.e. muons crossing the apparatus from below) are also searched; they are interpreted as produced by interactions inside the rock of energetic neutrinos ($2 \text{ GeV} \lesssim E_\nu \lesssim 100 \text{ TeV}$). Upgoing muons crossing the apparatus or stopping within it can be identified by using timing and tracking information; an excess in a particular direction in the angular distribution of these rare events could be an evidence of a stellar neutrino source. The first analysis of the upgoing muon data showed no evidence for such sources [MA93c].

The upgoing muon measured flux and angular distribution can be compared with some Monte Carlo calculations (e.g. [FR93]) which predict the number of expected events due to atmospheric neutrinos; a deficit or an excess of upgoing μ's could be interpreted as an evidence for a $\nu_\mu \leftrightarrow \nu_x$ oscillation ($x = e, \tau$ or sterile) in the matter or in vacuum. In ~ 2 years live time 74 ± 6 upgoing μ's were observed; the corresponding upgoing muon flux is $\phi \sim 10^{-12} \text{ cm}^{-2} \text{ s}^{-1} \text{ sr}^{-1}$. The observed number is $\approx 25\%$ lower than the expectations (101 ± 17) [MA95a]. When the experimental and theoretical uncertainties are taken into account, the probability that the observed number of events could differ from the central value of the expectation is 22%. The data are therefore compatible with the no-oscillation hypothesis.

The Weakly Interacting Massive Particles (*WIMP's*) can be indirectly detected by the observation of the upward-going muons. WIMP's trapped in the core of the celestial bodies, like the Sun or the Earth, can annihilate in ordinary particles (quarks and leptons), producing neutrinos; these neutrinos can induce upward muons. A search was performed in different angular windows around the direction of the core of the Earth and of the Sun; no statistically significant excess of muons was observed. The muon flux limits from the Sun and the Earth obtained by this search are $F_\odot \lesssim 2 \times 10^{-13}$ cm^{-2} s^{-1} and $F_\oplus \lesssim 6 \times 10^{-14}$ cm^{-2} s^{-1} [MA95b].

Finally, limits on the fluxes of strange (*"nuclearites"*) and exotic particles can be set at a level of $F \leq 10^{-14}$ cm^{-2} s^{-1} [MA92b].

3.3   General layout of the detector

The general layout of the MACRO detector [MA88a] is shown in fig. (3.3). It consist of six "supermodules" (*SM*), each one having a lower and an upper part (*"attico"*) and measuring $12 \times 12 \times 9$ m^3; the overall dimensions of the apparatus are $77 \times 12 \times 9$ m^3. Between each supermodule pair there is a 30 cm gap occupied by the mechanical support structure. The *"attico"* is hollow and contains the electronics for the experiment. The detector sides are designed by North, South, East and West; the North-South axis is not perfectly aligned with the true North direction, but the apparatus faces are referred as if it were. SM1 is the farthest north, SM6 the farthest south.

The main goal in the detector design was to obtain a large acceptance ($S\Omega \approx 9600$ m^2 sr) for an isotropic flux of penetrating particles and a large area ($S \approx 1000$ m^2) which ensures a high sensitivity in the search for rare phenomena in the cosmic radiation. The search for magnetic monopoles requires in addition a highly redundant detection system to obtain the best possible signature for such events.

Three independent and complementary methods are employed: three layers of scintillation counters (*"Bottom"*, **B**, *"Central"*, **C** and *"Top"*, **T**), 14 planes of limited streamer tubes (10 in the lower part and 4 in the *"attico"*) and a multi-layer plastic track-etch detector. The apparatus *"East"* (**E**) and *"West"* (**W**) faces are lined with 14 scintillation counters and the *"North"* (**N**) and *"South"* (**S**) faces with 7 scintillator tanks; the scintillation counters of these faces are vertically arranged and sandwiched between 6 streamer tube planes. An external layer of track-etch tiles is also present on the east and north walls of the detector. A Transition Radiation Detector prototype was recently installed above the *C* scintillation counter layer and is currently under testing.

The scintillation counters and the streamer tubes ensure the fast timing and energy loss measurements and the high resolution tracking of the penetrating particles, while the passive track-etch detector is particularly sensitive to slow

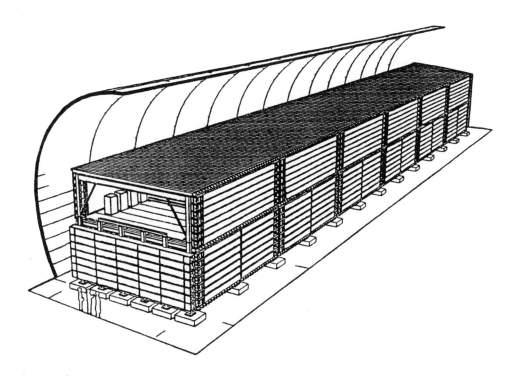

Figure 3.3: General layout of the MACRO detector.

highly-ionizing particles; the use of three different techniques ensures multiple signatures and cross-checks for rare events.

The large area and acceptance and the good tracking capabilities allow a sensitive search for astrophysical point sources, while the high-resolution timing is the fundamental tool in separating upward-going from downward-going particles at a level of $1/10^7$.

The occurrence time of the events in MACRO is measured by a Rb clock, whose $U.T.$ measurements are compared with that performed by the Italian Standard Time Institution; a 1 μs accuracy in the absolute time measurement is achieved. The timing information is carried through optical fibers to a network of slave clocks read by the acquisition system at each event.

In fig. (3.4) a building scheme of the first supermodule, showing the lower part and the *"attico"* mechanical skeleton, is presented. A description of the MACRO apparatus and particularly of the first supermodule is in [MA93a].

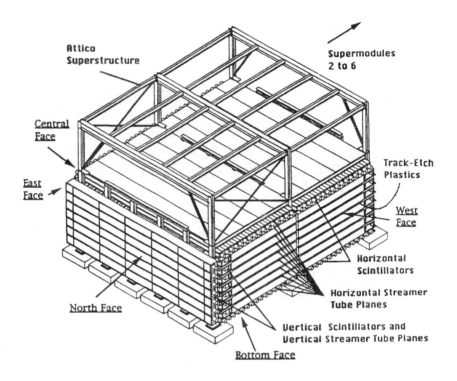

Figure 3.4: Building scheme of the first MACRO supermodule. The lower part, the detectors and the upper part skeleton are pointed out.

3.4   The scintillation counter system

The MACRO scintillation counter system consists of 294 horizontal counters (size $1200 \times 75 \times 20$ cm^3) and 182 vertical counters (size $1200 \times 25 \times 50$ cm^3); the active volume is $V_{hor} \approx 1120 \times 73.2 \times 19$ cm^3 in the horizontal and $V_{ver} \approx 1107 \times 21.7 \times 46.8$ cm^3 in the vertical counters. The remaining volume is occupied by two end compartments on the farthest counter sides, where the PMT's are located. The total active mass is ≈ 600 tonn.

The counters are referred to by a four letter designation, specifying the supermodule, the plane and the counter within that plane counting from North for horizontal counters and from the bottom for vertical counters; for instance, tank 2B03 is the third tank from North of the second supermodule, plane B. The counters are also indicated by a pure numerical code with the following rules: counters $1 - 96$ are for the B plane, counters $101 - 196$ for the C plane, counters $201 - 296$ for the T plane, counters $301 - 394$ for the W plane, counters $401 - 494$ for the E plane, counters $501 - 514$ for the N and S planes. In each plane the counters are grouped by 16 (horizontal planes) or by 14 (vertical planes); for instance, counters $117 - 132$ correspond to $2C01 - 2C16$ and counters $349 - 362$ to $4W01 - 4W14$. The left and right ends of the horizontal counters are defined watching the apparatus from South to North, that of the vertical counters watching the apparatus from West to East.

3.4.1 The liquid scintillator

The liquid scintillator consists of 96.4 % mineral oil and 3.6 % pseudocumene, mixed with 1.44 g/liter of *PPO* and 1.44 mg/liter of *bis-MSB*, used as wavelength shifters. The pseudocumene percentage was chosen to maximize the scintillator response for the ionizing particles [MA93a]. The ionizing energy is absorbed by the mineral oil molecules and transferred, by collisions, to the pseudocumene molecules; the ultraviolet light released by the pseudocumene is absorbed and re-emitted in the visible region by *PPO* and *bis-MSB*. The liquid scintillator properties are reported in the following table.

Table 3.1: Characteristics of the MACRO liquid scintillator.

Light Yield	56 % anthracene
Density	$0.87\ g\ cm^{-3}$
Attenuation length	> 26 m
Refractive index	≈ 1.48
Wavelength of maximum emission	420 nm

3.4.2 PMT's and light transport and collection

The scintillation light travels by counter internal reflections.

The counter inside is lined with teflon, which has a refractive index $n_{tef} \approx$ 1.33; this value, combined with that of the liquid, ensures a total reflection for incident angles $\theta \leq \theta_c = 25.6°$ relative to the walls; an air gap between the liquid level and the upper internal surface of the counters also produces total internal reflection at the air/scintillator interface (critical angle $\theta_c = 47.3°$).

The light is then collected at the two end compartments; they are filled with mineral oil (to ensure a good optical coupling) and separated from the liquid scintillator by PVC windows and contain two (horizontal counters) or one (vertical counters) 8 inch EMI PMT's (model D642KB) and one focusing mirror for each PMT[†]. The mirror profile was designed to increase the light collection efficiency (the PMT photocathodes cover $\approx$ 40 % of the scintillation counter transverse cross section) so that any critical light ray (i.e. a ray making

---

[†]The lower part of the first supermodule is equipped with R-1408 Hamamatsu PMT's.

a critical angle θ_c) coplanar with the revolution axis is reflected tangent to the photocatode surface; any ray coplanar with the mirror axis and with $\theta < \theta_c$ is guaranteed to reach the photocatode; the mirror design is therefore optimized for the light emitted by distant sources [LI88]. The optical scheme of the horizontal and vertical counters is shown in fig. (3.5).

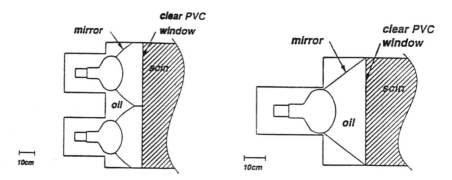

Figure 3.5: Optical scheme of the horizontal (left) and vertical (right) counters.

The PMT gains are set to have the peak of the single photoelectron pulse height distribution at 4 mV; the PMT working voltage (usually $\approx -(1200 \div 1700$ V)) is individually adjusted to achieve the requested gain. At this setting, a typical μ crossing a horizontal tank at the center produces a $1 \div 2$ V pulse. The signals coming from the two phototubes in a horizontal tank end are summed in a single signal carried by a 50 Ω cable; therefore we shall simply refer to a "single" PMT to indicate the pair of tubes at the same counter end.

The amount of the light collected on a PMT for a given energy deposit depends on the distance r of the emission point from the PMT; the light attenuation law can be simply described by the function:

$$f(r) = \alpha \, \exp\left(-\frac{r}{\lambda}\right) + \frac{\beta}{r^2} \tag{3.4.1}$$

where the first term (prevalent if the emission point is far from the PMT) represents the contribution of the light reflected by the box walls and the second term (prevalent if the emission point is close to the PMT) represents the contribution of the direct light. The counter attenuation length λ is mainly determined by the reflections at the counter walls and, to a smaller extent, by the scintillator attenuation length; its typical value is $\lambda \approx 12$ m. The two contributions in (3.4.1) are almost equal for $r \approx 1$ m.

The counter response, away from the PMT's, is practically uniform over the cross sectional area, but near the PMT's the probability to detect the scintillation light shows an enhancement closest to the tubes. At positions near a tube and to the side of the tank, a photon which is not emitted directly toward a PMT is likely to strike a wall and be absorbed or strike a reflector and be reflected back. This effect is present in both the horizontal and the vertical tanks in the last $1 \div 1.5$ m close to the PMT's and makes the event reconstruction and the energy resolution worse. Therefore, we sometimes define a "fiducial volume" excluding the last 1.5 m close to the tubes; this is usually done for calibration purposes (see ch. 6) and for studying the details of the radioactivity energy spectra (see ch. 5), when a good energy resolution is important.

3.4.3   Front End electronics and triggers

The signals from each counter end are fed through RG-58 (50 Ω) cables to linear *FAN-OUT's*, where 6 copies are made; from the *FAN-OUT's* the signals are distributed to the MACRO scintillator triggers. The dedicated stellar gravitational collapse trigger (*PHRASE*) will be discussed later; the others are two muon triggers (*ERP* and *CSPAM*) and two monopole triggers (*FMT* and *TOHM*).

Muon triggers

The *ERP* (Energy Reconstruction Processor) is a general muon trigger, which has also good capabilities for the stellar collapse neutrino detection. It operates with a high (~ 12 MeV) energy threshold for muons and a low (~ 7 MeV) energy threshold for stellar collapse events. A pre-trigger is generated by the coincidence of the signals from the two counter ends (after a 100 mV discrimination) within a 330 ns gate; the PMT signals are then integrated and compared with the contents of a look-up-table, set by software. The look-up-table contains, for any pair of signals, the event energy; the event is accepted if its energy is larger than the appropriate threshold. The look-up-table allows a trigger condition independent of the event position along the counter.

The *CSPAM* (the name is formed by the initials of the circuit designers) is a two plane muon trigger. The input is a fanned-in combination of the signals coming from 8 contiguous tanks. A plane pre-trigger condition is generated by the coincidence, within a 100 ns gate, of the fanned-in signals from both ends of a group of counters; the pulse height threshold is 200 mV. The trigger is given by two pre-triggers in two different planes within 1 μs.

Monopole triggers

The *FMT* (Fast Monopole Trigger) uses the same electronics of the *CSPAM*, but the coincidence between the pre-triggers is requested within 10 μs (instead of 1) and is vetoed by a *CSPAM* trigger within these 10 μs. The *FMT* is then sensitive to a β range from $\sim 5 \times 10^{-3}$ to $\sim 2 \times 10^{-2}$.

The *TOHM* (Time Over Half Maximum) is a slow monopole trigger, designed for detecting wide pulses of low amplitude or long trains of single photoelectrons (the PMT signals which would be produced by the passage of a slow monopole) and rejecting narrow and large pulses, caused by the radioactivity background and by the μ's. The *TOHM* produces a square pulse output with a width equal to the time over which the input pulse is larger than half of its maximum; its threshold is set to 2 mV, half the size of a single photoelectron pulse. The *TOHM* output is then sent to a "Leaky Integrator" digital circuit. The Leaky Integrator has two counting frequencies: a 66 MHz counting up frequency when the input (i.e. the *TOHM* signal) is high and a 1.5 MHz counting down frequency when the input is low. This arrangement produces a large number of counts when the input is a wide pulse or a concentrated train of photoelectrons and a small number of counts when the input is a narrow pulse or a series of well spaced pulses. A pre-trigger on one counter end is generated when the number of counts exceeds a preset threshold; a trigger is generated by a coincidence of two pre-triggers from both ends of a group of counters within a 20 μs gate. The slow monopole trigger is sensitive to a β range from $\sim 1.5 \times 10^{-4}$ to $\sim 2.5 \times 10^{-3}$ [MA93a].

Waveform digitizer systems

All these triggers can send a *STOP* signal to a system of waveform digitizers, made by a *Lecroy* 2262 (sampling frequency 50 MHz, time window length 6 μs) and a custom built system (sampling frequency 20 MHz, time window length 20 μs). The signals for the 50 MHz system are attenuated by a factor 5 and those for the 20 MHz system are amplified by a factor 10 to cover a different dynamic range.

3.4.4   Calibration systems

The correct behaviour and the stability of the scintillation counter system are periodically checked by means of an ultraviolet nitrogen ($\lambda = 337.1$ nm) laser. The laser output is passed through a computer controlled attenuator and an optical splitter; the light is carried to the tanks through 25 m long quartz fibers, mounted in a slotted PVC pipe.

The laser wavelength is well below the absorption band of the mineral oil and the pseudocumene, on the high edge of that of PPO and at the peak absorbency of bis-MSB. The PMT signals have different spectra and time behaviours for laser induced events and physical events (cosmic ray muons or natural background gamma rays); this is due to the fact that the ionizing radiation induces a scintillation light emission containing an amount of a slow component, practically absent in the laser case [BR79]. The time behaviours are shown in fig. (3.6).

There is also a LED-based calibration system, mainly used for calibration of TDC's and for simulating monopole pulse trains of various velocities and ionization strengths. The width and height of the LED pulses can be selected by using a programmable pulse generator.

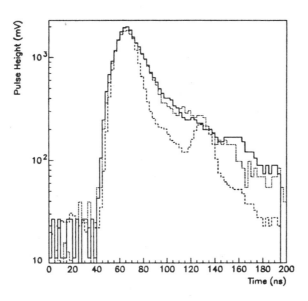

Figure 3.6: Normalized and superimposed pulses corresponding to laser light (dashed line), to a 4.4 MeV γ (dotted line) and to a cosmic ray μ (continuous line). The laser light pulse shape is different.

3.4.5   Counter resolution and time stability

The counter energy resolution can be interpolated by the form:

$$\frac{\sigma_E}{E} \approx 4\% + \frac{17\%}{\sqrt{E(\text{MeV})}} \tag{3.4.2}$$

where E is the energy deposited in the counter [BA90a]; the former term arises from the electronic digitization of the waveforms (see later), the latter from the photoelectron statistics; the number of photoelectrons is ≈ 20 for a 1 MeV energy release at the counter center. Several checks of (3.4.2) will be discussed later on.

The counter timing resolution is $\sigma_t \approx 500$ ps, corresponding to a position resolution $\sigma_x \approx 11$ cm at the cosmic ray energies.

The counter light yield as a function of time was measured by studying the muon energy loss distribution in the counters. This distribution (see ch. 5)

has a Landau peak at $E_{max} \approx 34$ MeV for vertical crossing; a decrease of the
scintillator response or a PMT gain loss would make this peak drift towards
lower energies; this effect can be software corrected by appropriate calibration
constants. The measured time variation is [BA95]:

$$\frac{\Delta Y}{Y} = (-1.3 \pm 0.6) \ \%/month \qquad (3.4.3)$$

The PMT gains are normally adjusted once per year.

3.5   The streamer tube system

The particle tracking in MACRO is based on the streamer tube system ([BA89b],
[MA93a], [MA95c]). Ten horizontal planes of tubes are contained in the lower
part of the apparatus. The first plane from the bottom is below the B counter
layer, the first from top is above the C counter layer and the remaining eight
are in the body of the detector and are separated by iron boxes filled with
the low-background crumbled Gran Sasso rock ($\rho \approx 2.71 \ g\,cm^{-3}$); because of
this arrangement, the minimum muon energy needed to vertically cross the
apparatus is $E_{min} \approx 2$ GeV. The 4 *"attico"* tube planes are mounted above
and below the scintillation counter T layer. The lateral planes are vertically
arranged, as the corresponding scintillation counter planes.

3.5.1   The streamer mode

The streamer mode [IA83] is an operating regime of gaseous detectors intermediate between the proportional and the Geiger mode.

The "streamer" is an ionization column, extending from the anode to the
cathode, which forms in case of very large amplifications (due to a strong
electric field between anode and cathode). When a charged particle ionizes
the gas, the electrons drift towards the wire, producing a charge avalanche;
the large amplification causes a big cloud of ion gas to form, which reinforces
the external field. The amplification in a streamer tube is so high that gas
ions and electrons can frequently recombine; the ultraviolet photons resulting
from the recombination cause a further gas ionization and produce secondary
charge avalanches within the ion cloud field. This process goes on and creates
the ionization column between anode and cathode. A quenching gas (usually
a hydrocarbon), added to the gas mixture, prevents ultraviolet photons from
liberating electrons from the cathode, causing afterpulses and a secondary
streamer.

3.5.2   Streamer tube structure and gas characteristics

The tube system has a modular structure; each module has a $1200 \times 3.2 \times$
25 cm^3 size and contains eight individual cells of a cross sectional area of

2.9×2.7 cm^2 ([BA86], [BA89c]). The horizontal planes use Cu/Be anode wires (diameter $\emptyset \approx 100\,\mu m$) and graphite cathode pick-up strips, placed at $26.5°$ with respect to the wires; this configuration provides a two-coordinate readout. The vertical planes use the wires only. The total number of wires for one supermodule is 5856 and the total number of electronic channels is 24000. The gas mixture consists of Helium (73 %) and n-pentane (27 %); the Helium was chosen to allow the exploitation of the Drell effect* for the slow monopole detection (The performances of the MACRO streamer tube system in the monopole search are discussed in [MA95c]). The total gas volume is ~ 200 m^3 and is flowed at a rate of 0.5 m$^3/hour$. The tube measured single rate shows a > 700 V wide plateau, upon which single ionization electrons are detected with a $\sim 20\,\%$ efficiency; the plateau center is at ≈ 4250 V ([IA83], [BA85]). The total streamer charge rises logarithmically with the ionization energy loss dE/dx for $dE/dx > (dE/dx)_{m.i.p.}$; the streamer tube ionization threshold is $(dE/dx)_{th} \sim 10^{-2} (dE/dx)_{m.i.p.}$. The maximum drift time in a tube is 600 ns; the corresponding temporal resolution is $\sigma_t \sim 150$ ns. After a hit, a tube is down for ~ 300 μs over a length ~ 5 mm.

3.5.3   Streamer tube electronics and trigger

The streamer tube readout is designed for recording the position (to reconstruct the particle track), the time (to measure the particle time of flight and velocity), the amount of the charge (to measure the particle energy loss) and the duration (to discriminate between slow and relativistic particles) of each streamer tube hit.

The wire front end electronics is based on 8-channel cards. Each wire has its own channel; the position information in a direction orthogonal to the wires is given by the address of the hit wire. The analog wire signals are summed and transmitted to an ADC/TDC system (the QTP); at the same time, the signals are discriminated and TTL shaped to form two pulses, of durations 10 μs and 550 μs. These pulses are sent to two shift registers, the Fast and the Slow Chain, which then contain the list of the streamer tube hits in the previous 10 μs and 550 μs respectively.

The strip system is read in a similar way: the signals are shaped to 14 μs and 580 μs to form a Fast and a Slow readout chain. The position information along the wire direction is extracted from the address of the fired strips and from the distribution of the induced charge.

The QTP

The time development and the total charge of the streamer tube signals are measured by the QTP (Charge and Time Processor). Each QTP channel

---

*The Drell effect [DR83] is an atomic level crossing induced, in some gases, by the high magnetic field associated with the passage of a monopole; this important effect was calculated for Hydrogen and Helium.

has, as input, 32 streamer tube wire channels, which correspond to 1 m of a horizontal plane; therefore, 12 QTP channels serve one whole horizontal plane. The time of the rising and falling edges of the signals is measured by a 6.6 MHz TDC; the corresponding precision is 150 ns, about the time resolution of the streamer tube system. The charge on each channel is measured by a 8-bit 20 MHz Flash ADC linear integrator; the charge range extends from 3 to 3000 pC, corresponding to a dynamic range of 20 mV to 1 V for relativistic particles and a pulse width range of 120 ns to 5 μs for a monopole-like pulse at the 20 mV threshold. The streamer signal arrival time, duration and charge are stored in a 63 byte, 640 μs cyclic memory.

Streamer tube triggers

Two triggers are based on the streamer tube system: the **Fast Particle Trigger** (FPT), which is mainly conceived for muons, and the **Slow Particle Trigger** (SPT), which is conceived for slow monopoles. The streamer tube trigger global sensitivity extends from $\beta = 10^{-4}$ to $\beta = 1$.

The FPT uses the Fast Chain pulses and takes advantage from the low noise ($R \sim 40$ Hz/m^2, dominated by the local radioactivity) of the streamer tube system; this low rate permits the definition of a trigger without need for tracking. The Fast Chain pulses in each 8-channel card are OR-ed and the resulting signal is OR-ed with that coming from the previous card. All the card signals of each plane are OR-ed in this way and the final signal is sent to a coincidence trigger circuit; this circuit performs a sampling of the signals, at a 3.3 MHz frequency, searching for appropriate combinations, like, for instance, one signal in each of 4 contiguous horizontal planes. A FPT is generated when one of these combinations is obtained.

The SPT uses the Slow Chain pulses and a 600 μs coincidence gate. To avoid a large number of accidental coincidences, the SPT condition is generated only if the streamer tube hits in the horizontal planes are in a temporal alignment defined by the detector geometry. For the vertical planes the time of flight between the inner and the outer planes is too short to make the time alignment sufficient to form a trigger, except for very low β particles. Hence, the vertical plane trigger requires in addition a spatial alignment between the hits.

3.5.4   Track reconstruction and streamer tube resolution

The particle tracks are reconstructed by searching for a set of (at least 4) aligned hits; these points are then linearly fitted to derive the track parameters. The position resolution is $\sigma_w \approx 1.1$ cm on the wire view and $\sigma_s \approx 1.6$ cm on the strip view; the corresponding intrinsic angular resolution is $\sigma_\theta \approx 0.2°$, but the global angular resolution is dominated by the multiple Coulomb scattering in the rock above the apparatus: $\sigma_\theta^{scat} \approx 1°$. This value was measured by using multiple muon showers and looking at the angular dispersion of their tracks relative to the shower axis ([MA93a], [MA93b]).

3.6   The track-etch and the transition radiation detector

The track-etch detector [MA88b] consists of 4 layers of *LEXAN* and 3 layers of *CR*39, respectively sensitive to fast ($\beta > 4 \times 10^{-3}$) and fast and slow ($4 \times 10^{-5} < \beta < 2 \times 10^{-4}$) monopoles, and one 1-mm-thick aluminum absorber, which stops the penetrating nuclear fragments. The detector is organized in "trains", each one containing 47 "wagons", 25×25 cm^2 in size; each wagon can be individually inserted and removed by sliding it on rails.

The passage of a heavily ionizing particle (a "monopole") causes some structural damage in the plastics; a chemical etching produces etch pits of equal size on both faces of the sheet; the pits are detected by the sheet microscope inspection. The position resolution of this detector is $\sigma_x \approx 100$ μm on each layer.

The Transition Radiation Detector (*TRD*) consists of 11 layers of a polyethylene radiator and 10 layers of proportional tubes (32 tubes for each layer), filled with Ar (90%) and CO_2 (10%). The present prototype is divided into two modules, each $6 \times 2 \times 2$ m^3 in size.

The expected peak for the transition radiation is at 11 KeV; its threshold corresponds to a Lorentz γ factor ~ 1000 or a muon energy $E_\mu > 100$ GeV. The *TRD* readout is based on the "cluster counting" technique. The transition radiation and the δ-ray production in the gas tubes induce ionization clusters on the *TRD* layers; the average number of detected clusters shows a poor dependence on the Lorentz γ factor for $E_\mu < 100$ GeV (below the transition radiation threshold) and a roughly linear dependence for 100 GeV $< E_\mu <$ 1 TeV; for $E_\mu > 1$ TeV the transition radiation saturates, giving a constant average number of clusters. The *TRD* can then measure the μ spectrum in the energy range 100 GeV $< E_\mu < 1$ TeV, which is of particular interest because at the MACRO depth the mean residual μ energy is ~ 240 GeV [MA94b].

3.7   Calorimetric properties of the MACRO counters at low energies

The MACRO scintillation counters have good calorimetric properties for low-energy charged particles; in fact, the minimum size of the counter, which roughly corresponds to the liquid level, is a factor ~ 4 larger than the range of a 10 MeV electron. Positrons from the reaction (2.1.1) would release nearly their whole energy in the counters. The expected efficiency for (2.1.1) positrons of a MACRO liquid scintillation counter is shown in fig. (3.7) for 3 different

energy thresholds: 5, 7 and 10 MeV. The rising edge of the curves agrees
with the experimental resolution (3.4.2); the plateau levels are $\approx 93\%$ for
$E_{th} = 5$ MeV, $\approx 90\%$ for $E_{th} = 7$ MeV and $\approx 86\%$ for $E_{th} = 10$ MeV.

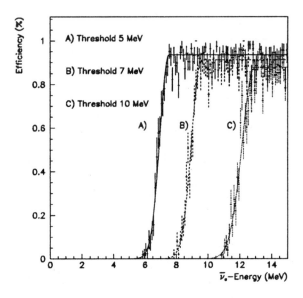

Figure 3.7: Expected efficiency of the MACRO liquid scintillation counters for
the $\bar{\nu}_e + p \rightarrow n + e^+$ reaction as a function of $\bar{\nu}_e$ energy for 3 different energy
thresholds. Calculations by the Monte Carlo method.

The calorimetric properties for the low-energy $\gamma's$ need a more detailed discussion, because low energy photons are detected in liquid scintillation counters
essentially via multiple Compton scatterings (the photoelectric effect and the
pair production give negligible contributions to the total photon cross section
for 0.1 MeV $< E_\gamma <$ 5 MeV). The average number of Compton diffusions ex-
perienced by a $\gamma_{2.2}$ in the counter was determined by the Monte Carlo method
[BA91a]: $\langle n_{Com} \rangle \approx 2.7$. In fig. (3.8) we show how 2 MeV photons are seen
in a MACRO counter if the number of Compton interactions is 1, 2, 3 and 4;
note that the mean energy increases and the line shape is better defined when
the number of Compton interactions grows; $n_{Com} = 3$ is sufficient for a line
shape good reconstruction. The photon energy is then well contained in the
counters.

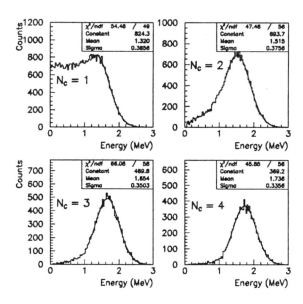

Figure 3.8: Reconstructed line shape of a 2 MeV γ in a MACRO liquid scintillation counter as a function of the number of Compton interactions (N_C) in the liquid.

The Compton scattering is highly directional and the Compton mean free path of a $\gamma_{2.2}$ in the liquid is ≈ 24 cm; the probability for observing a photon is rather low if the photon is emitted in a close-to-vertical direction, but rather high if the photon is emitted along the counter main axis. The counter has a high efficiency and good calorimetric properties for photons emitted along its main axis, but a low efficiency for photons emitted perpendicularly to this axis. Fig. (3.9) shows how 50000 $\gamma_{2.2}$'s, generated at the counter geometrical center, are seen in the scintillator if their initial directions are parallel to the main (top plots) or to the shortest (bottom plots) axis of the counter. In the main axis case, practically all photons are detected (i.e. experience at least one Compton interaction in the counter) and the average number of Compton interactions is ≈ 10.6; in the shortest axis case, $\approx 68\%$ of the photons are lost and the average number of Compton interactions is ≈ 1.6. A measurement of the $\gamma_{2.2}$ efficiency will be shown in ch. 6.

Two other interesting effects must be taken into account in the low-energy photon detection: the saturation of the scintillator response for electrons at

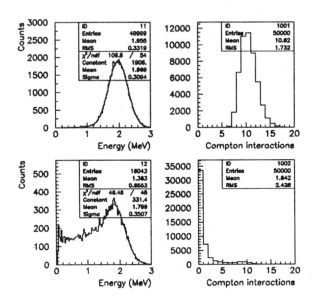

Figure 3.9: Response of a MACRO liquid scintillation counter for 2.2 MeV photons emitted along the main axis (top) or along the shortest axis (bottom) of the counter.

the end of their range and the energy spill-out from the counter.

The energy of the Compton electrons scattered by a $\gamma_{2.2}$ is ≤ 1.8 MeV (Compton edge); for electrons of such a low energy, the output of the liquid scintillator is not perfectly linear with the energy loss dE/dX of the electron [MI93], but is more correctly described by the Birks formula [BI51]

$$\frac{dL}{dX} = \frac{\frac{dE}{dX}}{1 + B\frac{dE}{dX}} \tag{3.7.1}$$

where dL/dX is the light yield (in units of energy/length) per unitary path length in the scintillator and B is the liquid "saturation constant". The constant B for the MACRO scintillator was measured [LI93] by using α and β particles; the result is

$$B = (13.3 \pm 1.4) \cdot 10^{-3} \text{ cm MeV}^{-1} \tag{3.7.2}$$

The saturation effect causes a $\approx 3.5\%$ reduction in the light yield at $E_{ph} = 2$ MeV.

A further effect is the counter energy leakage: photons and electrons can leave the counter without being completely absorbed; the observed electron and photon energy E_{obs} can therefore be lower than the photon initial energy E_{in}.

A plot of the visible energy in a MACRO counter vs the photon energy, as calculated by the Monte Carlo method, is shown in fig. (3.10) in double logarithmic scale; the triangles correspond to the visible energy when only the saturation is taken into account (the geometrical efficiency effects are removed by using a fictitious "infinite" counter, $100 \times 100 \times 100$ m^3 in size), the circles correspond to the visible energy when both the saturation and the geometrical effects are included.

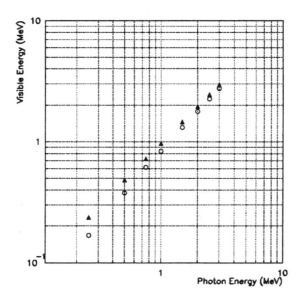

Figure 3.10: Monte Carlo calculation of the visible energy observed in a MACRO counter as a function of the photon energy.

In both cases, 6 different monochromatic sets of photons were generated in a counter (at the counter center in the first case and uniformly within the box in the second case); for each set, the reconstructed energy distribution was fitted, about the maximum, by using a gaussian shape; the corresponding visible energy is given by the gaussian peak position. Fig. (3.10) shows that the expected visible energy for a 2 MeV γ is ≈ 1.75 MeV; the scintillator saturation accounts for $\approx 30\%$ of the total effect.

The data in figure (3.10) were used to re-estimate the value of the Birks parameter B. By using the electron energy-range relation [MU76], a mean saturated and not-saturated energy loss were determined

$$\left.\frac{dE}{dx}\right|_{s,ns} = \frac{E_{s,ns}}{R(E)} \tag{3.7.3}$$

The pairs $(dE/dx|_s, dE/dx|_{ns})$ were fitted by the Birks formula

$$\left.\frac{dE}{dx}\right|_s = \frac{\left.\frac{dE}{dx}\right|_{ns}}{1 + B\left.\frac{dE}{dx}\right|_{ns}} \tag{3.7.4}$$

From the fit we obtained: $B = (15.2 \pm 1.6) \cdot 10^{-3}$ cm MeV^{-1}, in agreement with (3.7.2) within $1\ \sigma$.

Chapter 4

Stellar collapse trigger and data acquisition

4.1 General features of the data acquisition system

The acquisition system of the MACRO experiment was designed taking into account three main requirements [DA89]:

- the system must be modular and distributed in order to match the apparatus modularity;

- the system must be integrated in a network in order to allow an easy access from remote computers;

- the system must be largely based on commercial hardware and software products.

The system consists of a network (*Ethernet/DECNET*) of Digital computers, a VAX8200 (*"VXMACA"*) as file and network server and six μVaxII's. Three μVaxII's are dedicated to the acquisition of the data coming from three pairs of supermodules, while the remaining three are reserved for the stellar collapse data-taking. Each μVaxII is linked to a pair of CAMAC parallel branches via a CERN Fisher CAMAC System Crate. The use of a μVaxII network reduces the CAMAC cable length, since it is not necessary to have all the crates in the same location.

An extra μVaxII is used for testing and a group of VAXStations is reserved for monitoring and for displaying on-line pictures (*"Event Display"*). A VAX8200 (*"VXMACB"*) is dedicated to the data analysis. The MACRO Local Area Network (*LAN*) is connected to the area of the Gran Sasso laboratory (*LNGS LAN*) by a $2 - Mbit/s$ Decnet Bridge; from the external laboratory the wide area network (*WAN*) can be easily accessed. A general layout of the acquisition system is shown in fig. (4.1).

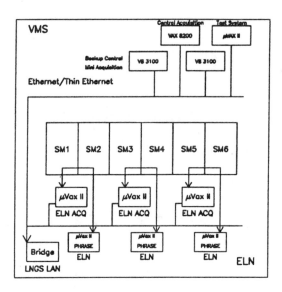

Figure 4.1: General layout of the MACRO acquisition system (after [MA93a]).

The μVaxII's run a *VAXELN* system. *VAXELN* is a Digital Software developed for real-time applications of the VAX processors; it has a fixed time response to external events and powerful message exchange facilities (useful for multi-job applications); its device drivers can be written using high-level languages (EPASCAL, C..). The CAMAC input/output is performed by using the standard CAMAC libraries, implemented in the *VAXELN* environment, and a CAMAC operation list.

The *VAXELN* section of the acquisition system consists of many jobs, running with different priorities; the most important are the four system jobs:

- *Event_In*, which reads the events;

- *Data_Reduction*, which performs a data reduction and zero suppression;

- *Event_Filter*, which makes an event pre-analysis based on the tracking;

- *Data_Spooler*, which sends the data to remote monitoring jobs (like the Event Display or the Supernova Monitor, which will be later discussed in detail).

The *VAXELN* facilities are used to exchange the data between different jobs. The maximum acquisition rate tolerable by the system is $\sim$ 90 Hz; the normal acquisition rate is $\lesssim$ 1 Hz, excluding the stellar collapse data.

VXMACA runs the standard *VAX/VMS* software; it manages a batch parent process, which controls the data acquisition, and many sub-processes (with different priorities), which collect the data and the alarm conditions coming from the μVaxII's, write the raw data on its internal disks (the mass storage capability is $\sim$ 3 Gbytes), route the commands to the μVaxII's and handle the general histogramming. The data between the μVaxII's and *VXMACA* can be exchanged at a rate $\lesssim$ 150 kbytes/s (40 kbytes/s if the data are written on disks). Other *VXMACA* multiple processes are the user interface for the run management (Consoles, On-line Histograms, Event Display ...). The data sharing among the processes is performed via mailboxes and global sections, the network services via the *DECNET* protocol. The system supports synchronous *DECNET* communications for directives coming from the *VAX/VMS* section which require appropriate acknowledgements (e.g. a completion message) and asynchronous *DECNET* communications for unsolicited data coming from the *VAXELN* section (e.g. alarm messages).

4.2 The stellar collapse acquisition system

The stellar collapse acquisition system is an independent subsection of the full MACRO acquisition system. It was designed following the common requests discussed above (modularity, use of commercial products etc.), but a great care was also devoted to improve the quality of the experiment as a supernova detector.

The first important point is that the stellar collapse data are collected by three dedicated μVaxII's; the stellar collapse acquisition system is therefore completely independent right up to the general data storage on the main computer; the common CAMAC list is not necessary. This avoids interruptions of the stellar collapse data-taking even if the general acquisition must be stopped, as frequently happens during the apparatus maintenance operations. The stellar collapse data acquisition can be managed by *VXMACA* or by a backup VAXStation3100.

The second point is the dead-time minimization. In case of an intense supernova neutrino burst ($\sim$ 1000 events in 10 s, uniformly distributed on the whole apparatus), a large dead-time would cause an event loss; the system must tolerate an acquisition rate up to $\sim$ 1 kHz. The dead-time reduction is obtained by a selective reading of the stellar collapse trigger circuits and by buffering the data.

Whenever a stellar collapse circuit needs to be read, it generates a Look At Me (*LAM*) signal; the *LAM* is used to identify which is the circuit to be read.

The *LAM* signals are handled by the Crate Controller (*CCA2*) and Branch Coupler (*BC*) modules. Every time a *LAM* is present, a $24 - bit$ Graded Lam Word (*GLW*) is built by the *CCA2*; the *GLW* contains the crate and station addresses of the module which generated the *LAM*. The OR of the *GLW's* from all crates in a branch is used to generate a Branch Demand (*BD*) and the OR of all *BD's* activates the *Event_In* process on the stellar collapse μVaxII's. The branches and crates are then sequentially scanned to search for the non-zero *GLW's* and identify the corresponding modules to be read; *Event_In* needs ≈ 4 ms to find and read a module which raised the *LAM*. Note that only the modules which generated the *LAM* are off during the reading operations.

After reading a module, *Event_In* resets its *LAM* and tests if a System Crate Demand (*SCD*) is present; if this is the case, other modules need to be read. The circuit that was just read is then not re-started, but its address is stored in a waiting list, while *Event_in* continues the reading cycle. When the *SCD* signal is no longer present, *Event_In* restarts the modules, scanning the waiting list in a FIFO order; at the end of the waiting list, *Event_In* is ready to receive new reading requests. This procedure avoids that, in case of a very high trigger rate per module, only the first modules in the scanning sequence are read.

The stellar collapse data are buffered in 16-event packets; this solution was adopted to reduce the *DECNET* loads during the communications between the stellar collapse μVaxII's and *VXMACA*. Each μVaxII has its own 16-event buffer; when the buffer is full, a lower priority spooler process sends the packet to *VXMACA*.

The fraction of lost events depends on the probability that the circuit, which should detect an event, is under reading or in the waiting list when the event occurs. In normal working conditions the dead-time is $t_{dead} \lesssim 0.004\% \; t_{tot}$; therefore no event is lost. The fraction of lost events in case of an intense neutrino burst was estimated by the Monte Carlo method simulating a 5000 event pulse within 10 s and following the reading sequence; the fraction of lost events was $< 0.3\%$.

The normal data acquisition rate is $\nu \sim 1 \div 1.5$ Hz for each stellar collapse μVaxII ($3 \div 4$ Hz on the whole apparatus); the disk occupancy is about 0.25 kbytes for each event, after the data reduction.

4.3   Requests for the stellar gravitational collapse electronics

In the search for ν's from stellar gravitational collapse one is faced with the problem of separating the real events from the background. This problem is less dramatic than for solar ν's, because of the pulsed character of the supernova signal.

The main background sources (which will be discussed in detail in the next ch.) are the cosmic ray muons and the natural radioactivity. The cosmic ray muons usually hit two or more counters releasing in each ~ 40 MeV; the time separation between the hits is < 270 ns, an upper limit set by the detector geometry; the apparatus crossing time for a vertical muon is ≈ 30 ns. The natural radioactivity γ's (here we do not distinguish between γ's from unstable radionuclides and γ's from neutron capture) hit one counter only; they give a signal similar to that of a (2.1.1) positron, but at an energy $\lesssim 10$ MeV. The γ spectrum is a rapidly decreasing function of the energy: the single counter trigger rate is ≈ 3 kHz for $E > 1$ MeV and ≈ 2 Hz for $E > 3.5$ MeV.

The trigger circuit must be able to reject a good fraction of the natural radioactivity background. Events with the same energy but at different longitudinal positions along a counter produce different PMT pulses (up to a factor 3) because of the light attenuation within the counter; this effect must be corrected if one wants a trigger depending only on the deposited energy. The energy threshold must be $\sim 5 \div 7$ MeV to efficiently reject the background and, at the same time, to be sensitive to a large fraction ($\gtrsim 85 \div 90\%$) of the $\bar{\nu}_e$ spectrum. The secondary $\gamma_{2.2}$ cannot be detected with such an energy threshold; the electronics must therefore be able to lower the threshold to a value $\sim 1 \div 1.5$ MeV after each primary event; the low threshold must be maintained for a time window much larger than the average neutron capture time (≈ 180 μs).

The PMT pulses for primary and secondary events must be recorded to measure their total charge (which is proportional to the deposited energy) and to extract their time information.

The event relative time must be measured with an accuracy (~ 1 ns), much lower than the muon transit time through the apparatus, to separate the e^+-like events from the cosmic ray background. Note that the mean separation between two ν events in a stellar collapse burst is $\sim 1 \div 10$ ms; the relative time accuracy is then perfectly adequate to reconstruct the ν burst temporal structure.

A correlated analysis between the data of different experiments which recorded ν's from the same supernova explosion is possible only if they measure the absolute time with an accuracy $\lesssim 1$ ms, much smaller than the ν burst duration. This accuracy is ensured in MACRO by the atomic clock; the stellar collapse trigger data must include the information coming from this device.

We now discuss how the requests for a stellar gravitational collapse trigger are satisfied by the dedicated *PHRASE* circuit.

4.4   The *PHRASE* circuit

PHRASE (fig. (4.2)) (**P**ulse **H**eight **R**ecorder **A**nd **S**ynchronous **E**ncoder) is a *CAMAC* circuit, developed as a specialized stellar gravitational collapse trigger.

The other circuit with good stellar gravitational collapse capabilities, the

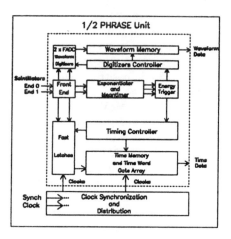

Figure 4.2: General logic of the *PHRASE* circuit.

ERP, was briefly described in the previous ch.; in this thesis only *PHRASE* data were used.

The signals from the PMT's at the counter ends are fed to the *PHRASE* circuits through the linear FAN-OUT's; each *PHRASE* module serves two counters. Between the FAN-OUT output and the *PHRASE* input there is a 0.5 factor attenuator.

PHRASE performs a group of functions:

- it gives a trigger condition (*"Primary Energy Trigger"*, *(PET)*), independent on the event longitudinal position along the counter, if the deposited energy is larger than a preset threshold $E_{th} \sim 7$ MeV;

- when a *PET* signal is present, *PHRASE* digitizes and stores the waveforms of the corresponding event, using 100 MHz Flash ADC's, whose dynamic range spans from 1 to 256 mV;

- it generates a *LAM* signal at PET; the *LAM* is used by the stellar collapse data acquisition system to identify the module to be read;

- after each *PET*, the circuit thresholds (one for each served counter) are lowered to ~ 1.5 MeV for a preset time interval (*"Special Time" ST*, usually $\approx 800\,\mu s$) to acquire secondary low-energy events, up to a maximum of 14. The secondary event trigger condition (*"Secondary Energy Trigger"*, *(SET))* is analogous to that of the primary events;

- *PHRASE* measures, with a 1.6 ns accuracy, the time of each event relative to the atomic clock standard time of the experiment; the time difference between the signals at the two counter ends is measured with the same accuracy. The synchronization between the *PHRASE* circuits is obtained by sending a common NIM signal to all the modules at each run start. The atomic clock is read by a CAMAC operation at each event.

The background rejection is performed in the following way: given the energy E and the longitudinal position x of an event in a liquid scintillation counter (x is measured from left to right, $x = 0$ at the counter left end and $x = L$ at the counter right end, L is the counter length), the corresponding pulse heights A_L and A_R on the left and right PMT's will be respectively

$$\begin{cases} A_L(E, x) = \alpha\ E\ \exp\left[-\frac{x}{\lambda}\right] \\ \\ A_R(E, x) = \alpha\ E\ \exp\left[-\left(\frac{L-x}{\lambda}\right)\right] \end{cases} \tag{4.4.1}$$

where λ is the counter light attenuation length and α is a constant. The light arrival times t_L and t_R on the PMT's will be

$$\begin{cases} t_L(x) = t_{MT} + \left(\frac{x-L/2}{v}\right) \\ \\ t_R(x) = t_{MT} - \left(\frac{x-L/2}{v}\right) \end{cases} \tag{4.4.2}$$

where $T_{MT} = (t_L + t_R)/2$ is the *"Mean Time"* and $v \approx 0.2\,\text{m}/\,\text{ns}$ is the speed of light along the counter. If two exponential tails are electronically added to (4.4.1), the new pulse heights S_L and S_R are

$$\begin{cases} S_L(E, x, t) \propto\ E\ \exp\left[-\frac{x}{\lambda}\right]\ \exp\left[-\left(\frac{t-t_L}{\tau}\right)\right]\ \Theta(t - max\{t_L, t_R\}) \\ \\ S_R(E, x, t) \propto\ E\ \exp\left[-\left(\frac{L-x}{\lambda}\right)\right]\ \exp\left[-\left(\frac{t-t_R}{\tau}\right)\right]\ \Theta(t - max\{t_L, t_R\}) \end{cases} \tag{4.4.3}$$

where τ is the tail time constant and $\Theta(x)$ is the Heaviside function ($\Theta(x) = 1$ if $x > 0$ and $\Theta(x) = 0$ if $x \le 0$). Choosing $\tau = \lambda/v \approx 60\,\text{ns}$ and taking into account (4.4.2) we obtain

$$\begin{cases} S_L(E, x, t) \propto\ E\ \exp\left[-\left(\frac{t-t_{MT}}{\tau}\right)\right]\ \Theta(t - max\{t_L, t_R\}) \\ \\ S_R(E, x, t) \propto\ E\ \exp\left[-\left(\frac{t-t_{MT}}{\tau}\right)\right]\ \Theta(t - max\{t_L, t_R\}) \end{cases} \tag{4.4.4}$$

In (4.4.4) every non-uniformity due to the event position has been canceled because of the particular ad-hoc choice of τ; therefore, at a fixed time t^* ($> t_{MT}$), the pulse heights S_L and S_R are only functions of the deposited energy E and depend linearly on E. The condition $t^* > t_{MT}$ can be always satisfied choosing $t^* = 40$ ns (which is larger than half the maximum time needed for a light signal to longitudinally cross a counter: $t_{cross} = \frac{L}{2v} \approx 30 \ ns^\ddagger$). The trigger condition is generated when both the pulse heights are above the threshold, at the time t^*.

When a PET occurs, $PHRASE$ Flash ADC's digitize and record the event waveforms, using the techniques developed for the CERN circuit $GATHER$ [BO84]. The Flash ADC response is linear in a dynamic range corresponding to small energy deposits. From the experimental resolution we estimated (see ch. 6) that a 1 MeV energy deposit at the counter center produces ~ 20 $photoelectrons$; the PMT gains are set to 4 mV/phel; the attenuator in front of the $PHRASE$ circuits reduces this value by a factor 2. Taking into account the time width of the pulse, we obtain that the linearity limit (256 mV) corresponds to an energy release $\Delta E \sim 15$ MeV at the counter center. For input signals < 256 mV, the digitized waveform area is proportional to the energy deposited. For input signals > 256 mV the Flash ADC's saturate, but the width of the digitized waveforms continues to increase; its area also increases (even if not linearly) and a unique relation between the integral of the digitized pulse and the energy deposit is still present. The non-linearity can then be corrected and a wide linear energy scale can be determined (see § (6.1)).

Incidentally, the dynamic range can also be extended using the fact that the optical reflections on the PMT mirrors produce delayed pulses, whose height is about $5 \div 10 \%$ of that of the main pulse; an upper limit on the deposited energy (independent on the algorithm used to linearize the energy scale) can then be determined. This technique was applied in the search for medium speed monopoles ($10^{-3} < \beta_{mon} < 10^{-1}$) using $PHRASE$ data [BA92a].

4.5   Off-line position and energy reconstruction

The position of an event along a liquid scintillation counter can be reconstructed using the difference in the light arrival time at the two counter ends. If t_L and t_R are the arrival times on the left and right end and v is the light velocity in the counter, the event position x is given by:

$$x = \frac{v \ (t_L - t_R) + L}{2} \qquad (4.5.1)$$

$\ddagger$The time origin in these calculations is the minimum between t_L and t_R. For instance, suppose $t_L < t_R$: then $t_L = 0$ and $t_{MT} = \frac{t_R}{2} = \frac{1}{2}(\frac{x}{v} + t_{MT})$, or $t_{MT} = \frac{x}{v} \le \frac{L}{2v}$

($x = L/2$ at the counter center ($t_I = t_R$)). The arrival time information is supplied by a pair of fixed threshold discriminators (one for each end), whose thresholds are continuously compared (with the *PHRASE* sampling time $t_s = 1.6$ ns) with the level of the signals. A lower precision time information is related to the pulse average time. If these two measurements agree within 5 ns, the more precise discriminator information is retained; otherwise, the pulse information is assumed to be correct, since the discriminator time is less reliable for small, irregular pulses.

The energy reconstruction is based on a weighted mean of the pulse amplitudes on the two counter ends.

The measured pulse amplitudes are converted by the linearizing algorithm (see § (6.1)) in "linearized" pulse amplitudes A_L and A_R; then, A_L and A_R are multiplied by appropriate factors which balance the different PMT gains and the *PHRASE* channel amplifications; a further multiplicative factor converts the arbitrary pulse amplitude scale in an absolute energy scale.

Using the reconstructed position x, two light attenuation factors, S_L and S_R, are determined

$$\begin{cases} S_L(x) = \exp\left[-\left(\frac{x-L/2}{\lambda}\right)\right] \\ S_R(x) = \exp\left[-\left(\frac{L/2-x}{\lambda}\right)\right] \end{cases} \tag{4.5.2}$$

(S_L and S_R are normalized to unit at the counter center ($x = L/2$)). The reconstructed energies E_L and E_R on the two counter ends are given by:

$$\begin{cases} E_L = A_L/S_L(x) \\ E_R = A_R/S_R(x) \end{cases} \tag{4.5.3}$$

The amplitudes A_L and A_R are generally different; then, they have also different uncertainties ΔA_L and ΔA_R due to the photoelectron number fluctuations. The corresponding uncertainties $\Delta E_{L,R}$ on $E_{L,R}$ can be estimated by x, $A_{L,R}$ and $\Delta A_{L,R}$ and the formula (4.5.3). ΔE_L and ΔE_R are the weighting factors for E_L and E_R in computing the energy E; a larger weight is attributed to the pulse which has smaller fluctuations. The energy E is given by:

$$E = \frac{\frac{E_L}{\Delta E_L^2} + \frac{E_R}{\Delta E_R^2}}{\frac{1}{\Delta E_L^2} + \frac{1}{\Delta E_R^2}} \tag{4.5.4}$$

In case of balanced uncertainties ($\Delta E_L = \Delta E_R$), (4.5.4) becomes simply

$$E = \frac{E_L + E_R}{2} \tag{4.5.5}$$

as expected.

4.6   Experimental and environmental conditions and data quality checks

The continuity of the data acquisition is a fundamental requirement for a stellar gravitational collapse experiment; the data quality must also be continuously checked.

The environment conditions (temperature and humidity) are stabilized in the MACRO tunnel by an air ventilation and conditioning system. A good air ventilation is also useful to remove the ^{222}Rn, which diffuses from the rock and concrete in the air; being a heavy gas, it concentrates in bad-ventilated environments. ^{222}Rn is one of the main sources of natural radioactivity and an abnormal ^{222}Rn concentration can perturb the data quality and be a health hazard. ^{222}Rn is an α-emitter; its decay products ("Rn daughters") are also α and γ emitters. The Rn daughter γ-ray's have energies $1.7 \div 2.2$ MeV, the energy range where we expect the delayed $\gamma_{2.2}$ signal. The ^{222}Rn concentration is permanently monitored (e.g. [AR94]); it varied considerably in the past, because of the construction and improvement of the ventilation system. Presently (spring 1995) the ^{222}Rn concentration in the gallery is $\approx 40\,Bq/m^3$, a factor 3 below the "minimum attention value" ($135\,Bq/m^3$) fixed by the U.S. Environment Protection Agency (EPA).

Disturbances on the electric power lines are an important source of noise for the stellar gravitational collapse trigger; they can cause spurious triggers by suddenly increasing the experiment counting rate (for instance: by changing the gain of a group of PMT's). Power glitches can also cause a synchronization loss of some PHRASE modules; in this case, some muon events can be erroneously flagged as positron-like events. The power line voltage and frequency are stabilized within 2 % by a 250 kW commercial stabilizer group; the residual anomalies are automatically timed and recorded by a computer controlled power line monitor ("DRANETZ 646-3"). In ch. 7 some examples of spurious clusters of events showing a strong time correlation with electric disturbances will be discussed.

Finally, the apparatus correct behaviour can be rapidly and efficiently checked by continuously monitoring quantities like the time stability and the uniformity of the counting rate on the whole scintillation counter system. This check is performed by the program LSCM ("Liquid Scintillation Counter Monitor"), a spy job, which works as a low-priority subprocess of the normal acquisition system; LSCM identifies and diagnoses the apparatus anomalies.

This program classifies the primary events into 2 categories:

- coincidences: events which cause hits in two or more counters (within 320 ns) or one counter and the streamer tube system within 5 μs (this time window takes into account the streamer signal time formation ($\sim$ 0.5 μs) and the length of the cables which feed this signal from the tracking electronics to a group of dedicated PHRASE modules);

- *singles*: events which hit one scintillation counter only.

The *single* event rate for each supermodule is computed using the last 150 observed events (*"running rate"*); the multiplicity (number of events within the *coincidence* interval) of each *coincidence* is also recorded and the running rate of the *coincidence* multiplicity is calculated.[‡] The program computes also the time averaged values (during the previous and the current period (1 hour)) of the following quantities:

- the liquid scintillation counter *single* rate (~ 0.5 Hz per supermodule);

- the liquid scintillation counter (*LSC*) *coincidence* multiplicity; since cosmic ray μ's usually hit ≥ 3 counters, this number must be $\gtrsim 3$, when the apparatus is well behaving. The streamer tube signal is not included;

- the liquid scintillation counter *coincidence* rate (total and per supermodule). The number of hit counters is computed taking into account the coincidence multiplicity, i.e. when the multiplicity of a *coincidence* event is M, the number of hit counters is increased by M. The expected total rate can be estimated as follows: the μ flux at the MACRO depth is $\phi_\mu \sim 1 \; m^{-2} \; h^{-1}$, the apparatus surface is $S \sim 1000$ m^2 and the LSC average *coincidence* multiplicity is $\gtrsim 3$. The μ rate is then $R_\mu \approx 0.26$ Hz and the LSC *coincidence* rate is $\approx 0.26 \times 3$ Hz ≈ 0.8 Hz;

- the scintillation counter & streamer tube (*ST*) *coincidence* rate. The number of LSC & ST *coincidence* events is increased by one whenever a temporal and spatial coincidence between the LSC and the ST triggers is present, regardless of the number of hit counters (i.e. the LSC coincidence multiplicity) of the event. The expected rate is then determined by the μ flux and by the apparatus surface and is about one third of the previous one (in normal working conditions both the LSC and the ST systems have an almost 100 % trigger efficiency for cosmic ray muons);

- the streamer tube rate (total and per μVaxII). Note that when a cosmic ray muon causes hits in supermodules served by different μVaxII's, a muon trigger signal is sent by each of these μVaxII's to the dedicated *PHRASE* circuits. The recorded streamer tube rate is therefore significantly larger than the LSC & ST coincidence rate. Since the μ's preferentially come from the apparatus South-North direction, the geometrical acceptance and the streamer tube trigger efficiency are maxima for the second μVaxII (supermodules 3 and 4), which has then a larger trigger rate than the two others.

[‡]An event recorded only by the tracking system is flagged as a "multiplicity one" *coincidence* and is indicated by a point along the *"ST Events"* line in the *LSCM* display.

All these quantities are shown in a graphical display, as in fig. (4.6); the time averaged values are reported on the right of the picture.

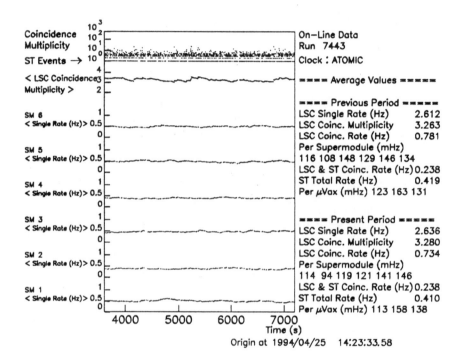

Figure 4.3: Graphical display of the *LSCM* program.

If some anomalous condition is detected, the program displays messages and warnings and alerts the shiftworkers by means of acoustic signals; a corresponding message is also sent when the correct working conditions are recovered. In particular, *LSCM* can recognize a synchronization loss of some *PHRASE* modules (from a too low *coincidence* rate), or an anomaly of the streamer tube system, or a strange, abnormally high, rate of one or more counters ("firing counters") etc. The delay between the anomalies and the corresponding warnings varies from ~ 10 s to $\sim 30^m$; generally, more serious is the anomaly, faster is the alarm. Simple interventions by the shiftworkers are sufficient to restore the normal working conditions, at least in a part of the apparatus.

Chapter 5

Background sources

In an underground experiment like MACRO the main background sources are the natural radioactivity and the penetrating cosmic ray μ's. The radioactive elements present in the Gran Sasso rock and in the building materials emit mainly low-energy ($\lesssim 2.5$ MeV) γ-lines and also neutrons and high-energy ($\lesssim 7$ MeV) γ-lines from spontaneous fissions. The cosmic ray μ's produce, by collisions and captures, neutrons and spallation products. The background in the liquid scintillators due to each of these sources can be partially rejected by appropriate techniques; the residual noise limits the experiment sensitivity.

5.1 Natural radioactivity

The natural radioactivity background in MACRO has been extensively studied ([BA90b], [BA91a], [BA91d], [BA93a], [BA95]). The scintillation counters, which have a large surface/volume ratio and no passive shielding, are particularly vulnerable to the external background. Since the natural radioactivity energy spectrum rapidly drops after ≈ 2.5 MeV, only the signals detected within the secondary event time window ($E_{th} \sim 1$ MeV, $\Delta t \approx 800$ μs) get contributions from this background.

The natural radioactivity originates from the decay of radioactive isotopes present in the Gran Sasso rock, in the concrete used in the laboratory and in the building materials of the experiment ([FI85], [BE89]). Gamma rays are emitted by the elements of the ^{238}U and ^{232}Th chains or by ^{40}K and are partially degraded by re-interactions in the rock and in the concrete, before being detected by the liquid scintillation counters. The local amount of Radon is important in determining the concentrations of the various radioactive species, since the ^{222}Rn diffusion in the air destroys the ^{238}U chain secular equilibrium.

Large, high-resolution, Ge ([AR92a], [AR92b]) and NaI [BA93a] detectors were used for measuring the natural background; in fig. (5.1) a radioactivity energy spectrum measured by a cylindrical ($\phi = 3''$, $h = 3''$) Sodium Iodide detector is shown.

The main features of this spectrum are the following:

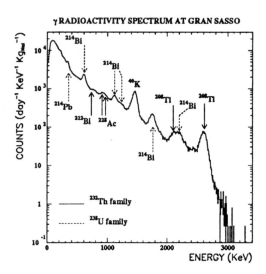

Figure 5.1: Radioactivity background energy spectrum measured by a $NaI(Tl)$ detector.

- a rapidly decreasing, approximately exponential, continuum, due to the overlap of the low energy tails of many, partially degraded, lines;

- a group of lines coming out of the continuum; they can be identified because of the detector good energy resolution. The line at 1.46 MeV is the ^{40}K decay line, the one at 2.61 MeV is the ^{208}Tl line, the others are mostly lines emitted by ^{214}Bi (an element of the ^{238}U family). A list of energies and approximate intensities (based on the measurements reported in [FI85]) is given in table (5.1);

- a sudden drop (by a factor $\approx 50 \div 100$) at energies higher than the 2.61 MeV ^{208}Tl line; higher energy contributions come from rare ^{214}Bi-lines and from γ-lines due to neutron capture by the I of the NaI detector.

When the radioactivity energy spectrum is measured by a large liquid scintillation counter, the individual lines can no longer be singled out, because of the much poorer energy resolution of this detector; liquid scintillation counters have a lower number of photoelectrons/ MeV than NaI detectors. The radioactivity events are reconstructed in energy and position by using the PMT time

Table 5.1: γ-Lines in the Gran Sasso Rock and Concrete.

Element	Line Number	Energy (MeV)	Intensity (A.U.)
^{208}Tl	1	0.583	0.12
^{214}Bi	2	0.609	1.00
^{214}Bi	3	0.768	0.11
^{228}Ac	4	0.911	0.13
^{214}Bi	5	0.934	0.07
^{228}Ac	6	0.960	0.08
^{214}Bi	7	1.120	0.23
^{214}Bi	8	1.238	0.04
^{214}Bi	9	1.281	0.03
^{214}Bi	10	1.378	0.09
^{214}Bi	11	1.401	0.05
^{214}Bi	12	1.408	0.03
^{40}K	13	1.461	0.49
^{214}Bi	14	1.509	0.05
^{214}Bi	15	1.729	0.07
^{214}Bi	16	1.764	0.16
^{214}Bi	17	1.847	0.04
^{214}Bi	18	2.204	0.11
^{214}Bi	19	2.448	0.03
^{208}Tl	20	2.614	0.15

and charge information, as in § (4.5). The measured radioactivity spectrum depends on the electronic detection efficiency and acquisition threshold; in the case of MACRO counters and *PHRASE* circuits, the electronic efficiency $\epsilon(E)$ as a function of the energy E can be approximated by the gaussian integral function

$$\epsilon(E) = freq\left(\frac{E - E_{th}}{\sigma_{th}}\right) \qquad (5.1.1)$$

where E_{th} is the *PHRASE* acquisition threshold, σ_{th} the resolution at $E = E_{th}$ and

$$freq(x) = \frac{1}{\sqrt{2\pi}} \int_{-\infty}^{x} \exp\left(-\frac{y^2}{2}\right) dy \qquad (5.1.2)$$

Examples of high statistics radioactivity energy spectra measured by a horizontal (left) and by a vertical (right) counter are shown in fig. (5.2); a software cut was applied to eliminate data corresponding to the 1.5 m region close to the PMT's.

Both spectra exhibit a characteristic shape change at the ^{208}Tl line; they

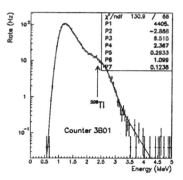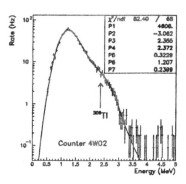

Figure 5.2: Radioactivity background differential energy spectra measured in a horizontal (left) and a vertical (right) liquid scintillation counter.

can be fitted by the combination of an exponential and a gaussian form, folded with the response function (5.1.1), as in fig. (5.2). The energy scales were set (see ch. 6) to position the Tl-line (mean of the gaussian curve, parameter P4 in fig. (5.2)) at 2.37 MeV (instead of 2.61 MeV) for taking into account the saturation of the liquid scintillator and the counter energy-leakage, as discussed in § (3.7). The counter energy resolution at the Tl-line position is measured by the standard deviation of the gaussian curve (parameter P5); the ratio $\sigma_E/E \approx 13 \div 14\,\%$ at $E = 2.61$ MeV for horizontal counters is in agreement with the expression (3.4.2). Vertical counters, with one PMT only at each end, have a lower light yield and therefore a poorer energy resolution ($\sigma_E/E \sim 20\,\%$ at $E = 2.61$ MeV).

Far from the energy threshold, counters belonging to the same layer have very similar differential energy spectra (excluding some pathological cases, for example counters having one dead PMT); in fig. (5.3) the differential energy spectra of 16 counters of the SM 3, layer C and 12 counters of the SM 4, layer W are shown superimposed.

The similarity among the energy spectra looks particularly good for horizontal counters, while the vertical counter spectra show less uniform and regular characteristics.

It must be stressed that the spectra become equal only if the energy calibration (which will be later described in detail) is accurate. Suppose ΔE is a shift in the energy scale; the corresponding variations $\Delta (dR/dE)$, in the differential, and ΔR, in the integral rate, can be estimated approximating the decreasing portion of the radioactivity energy spectrum with an exponential form of slope λ (this approximation corresponds to neglect the contribution of

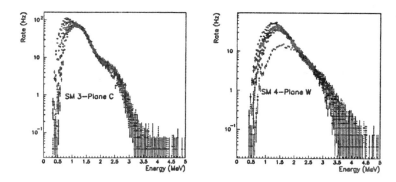

Figure 5.3: Superimposition of the differential energy spectra of 16 horizontal counters (left) and 12 vertical counters (right).

the Tl-line, which introduces a small correction at $E \lesssim 2$ MeV). The result is:

$$\left| \frac{\Delta \left(dR/dE \right)}{dR/dE} \right| \simeq \lambda \, \Delta E \qquad\qquad \left| \frac{\Delta R}{R} \right| \simeq \lambda \, \Delta E$$

Being $\lambda \sim 3$ MeV^{-1} we have $|(\Delta \left(dR/dE \right)) / \left(dR/dE \right)| \approx |\Delta R/R| \approx 30\,\%$ for $\Delta E = 0.1$ MeV. A small change in the energy scale can cause large variations in the differential and integral rates, because the energy spectra are very steep around $E \sim 1 \div 2$ MeV.

Horizontal counters belonging to different layers may be exposed to different natural radioactivity fluxes; actually, the B counters, located close to the hall floor[‡], have larger integral rates (about a factor 2) than the C counters, which are far from the floor and partially shielded by the apparatus mechanical structure [BA93a]; the T counters have larger integral rates than the C ones, because they are exposed to the radioactivity γ's coming from the hall ceiling.

Various effects must be taken into account in comparing the integral rates of horizontal and vertical counters: the different exposure, the different geometrical sizes (the horizontal counter volume is ≈ 1.5 times that of the vertical counters) and the vertical counter poorer energy resolution, which causes a broadening of the energy spectrum. In fig. (5.4) the integral rates of the $3B01$ and $4W02$ counters are shown; since they are both located close to the floor, they are exposed to a similar radioactivity flux. For low energy cuts ($E_{cut} = 1.5 \div 2$ MeV), the ratio $R_{hor}/R_{ver} \sim 1.5$ is in agreement with what expected from geometrical considerations; for higher cuts, the resolution ef-

---

[‡]The radioactivity content of the concrete is higher than that of the Gran Sasso rock.

fects play a more important role and the ratio decreases to $R_{hor}/R_{ver} \sim 1$ at $E_{cut} > 3$ MeV.

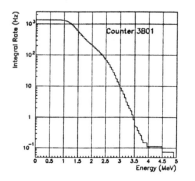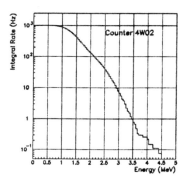

Figure 5.4: Radioactivity background integral energy spectra measured in the horizontal (left) and the vertical (right) liquid scintillation counters of figure (5.2).

Finally, the longitudinal position distribution of the radioactivity events in a MACRO counter ($1B14$) is shown in fig. (5.5); three different energy cuts (1, 1.5 and 2 MeV) were applied. The vertical lines represent the counter limits.

One can see that the position distributions along the counter become more uniform when the energy threshold is increased. In the lowest energy distribution (top of figure) the number of events is larger on the right half ($x > 560$ cm) and increases going from the counter ends towards the center; for $E > 1.5$ MeV, the difference between the left and right half is reduced, but there is still an excess of events close to the counter center; finally, for $E > 2$ MeV the distribution is reasonably uniform, with some fluctuations due to the lower statistics.

The shapes of the distributions are determined by two effects:

- the hardware selections operated by the discriminators and the energy trigger;

- the photoelectron number fluctuations.

At event energies $E \sim 1$ MeV the PMT pulses are very small. Suppose for a moment that no fluctuation occurs and consider events of equal energies in different longitudinal positions. The events near the counter center produce similar pulse heights on the PMT's at the opposite counter ends; such pulse

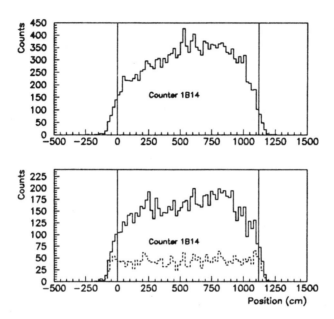

Figure 5.5: Radioactivity background position distributions measured in the counter $1B14$ with a 1 MeV (top), a 1.5 MeV (bottom, continuous line) and a 2 MeV (bottom, dashed line) energy cut.

heights are usually both larger than the discriminator thresholds and the energy trigger selection applies. On the contrary, the events near one counter end produce, on the far PMT, a pulse which can be up to a factor 3 smaller than that on the close PMT; the far pulse can then be frequently lower than the corresponding discriminator threshold. In this case, the event is not accepted, even if the energy deposit is larger than the ET threshold. This causes the "plum-cake" shape of the distributions with lower energy cuts; increasing the energy threshold, the effect becomes less important, since, at higher light levels, only the events very close to one counter end can produce, on the far PMT, a pulse below the discriminator threshold.

The left-right asymmetries are caused by the photoelectron number fluctuations. The two PMT's of the counter shown in fig. (5.5) have different responses at fixed light levels: the number of photoelectrons/ MeV on the right PMT ($x = 1120$ cm) is ≈ 1.2 times larger than that on the left PMT

$(x = 0)^{\S}$; then, the fluctuations are larger on the (worse) left PMT. Now consider two events of the same energy, one close to the left PMT and one to the right PMT; the energy deposit is supposed above the ET threshold. In both cases the event will be accepted if the pulse on the far PMT is larger than the corresponding discriminator threshold; this happens more frequently when the fluctuations are larger, i.e. near the left PMT. An excess of events is therefore expected close to the right PMT, as observed.

The photoelectron number fluctuations also deteriorate the quality of the time information near the counter ends; this causes the longitudinal position distributions to extend beyond the counter limits.

5.1.1   Effects on the trigger rates of new materials added to the detector

Each time some new material is added to the MACRO apparatus, it must be rapidly checked if this causes a background increase in the liquid scintillation counters.

During the winter 1994, a 5 cm thick firebreak slab was inserted below the T counters and a double layer (1.2 cm of plasterboard and 0.5 cm of corrugated iron) floor was laid over the T plane. The effect of these materials on the radioactivity background rates in the horizontal counters was studied by using both the measurements of the plasterboard and firebreak layer activities and the PHRASE data [BA95]. The results are summarized:

- the B counters were practically unaffected. This corresponds to what one expects, since the B plane is shielded from the radioactivity coming from the *"attico"* by the C plane and by the interspaced absorbers;

- the C counters showed a modest counting rate increase ($\Delta R/R \sim 15\%$) after the firebreak layer mounting and no variation after the floor mounting. These counters are exposed to the radioactivity coming from the firebreak layer and are shielded from the radioactivity coming from the *"attico"* floor by the T plane;

- the T counters showed a modest counting rate increase ($\Delta R/R \sim 18\%$) after the firebreak layer mounting and a similar decrease ($\Delta R/R \sim -20\%$) after the floor installation. They are exposed to both the radioactivity sources (the firebreak layer and the plasterboard), but the plasterboard radioactivity is ~ 3 times lower than the firebreak layer radioactivity and the iron slab within the floor is a good shield for the natural background coming from the ceiling concrete.

$\S$We will discuss the techniques for measuring and balancing such different responses in the next ch.

In fig. (5.6) the comparison between the integral counting rates after and before the firebreak layer installation (full circles) and after and before the floor mounting (open circles) is shown for a sample of 25 T counters, selected by requiring regular spectral shapes, a good statistics and a reliable energy calibration; the quoted errors are only statistical. The counters 247 and 250

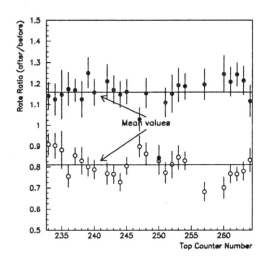

Figure 5.6: Comparison of the integral counting rates on T counters after and before the firebreak layer installation (full circles) and after and before the "attico" floor mounting (open circles).

have anomalous ratios due to unusual gain variations ($\Delta G/G \sim -2\%/month$ for counter 247 and $\Delta G/G \sim -2.5\%/month$ for counter 250, see (3.4.3) for comparison).

The conclusion is that these new materials did not deteriorate the quality of the stellar collapse data.

5.2 γ's from neutron capture

Fast and thermal neutrons are a further source of γ rays, with energies reaching the nucleon binding energy. Neutrons are produced by ^{238}U spontaneous fission and by cosmic ray spallations on materials close and within the experiment. They can be moderated and captured in the scintillator and in the other materials within the detector, as the Chlorine of the PVC boxes or the

Iron of the support structure. The neutron capture leaves a nucleus in an excited state; its de-excitation proceeds via multiple photon emission with a total energy release $E_{tot} \approx 8$ MeV.

The neutron flux in the laboratory was measured using detectors based on neutron capture on Cd ([BE85], [RI88]) and B [BE89]; the most recent results are

$$\phi_n = \begin{cases} (1.07 \ \pm \ 0.05) \cdot 10^{-6} \ \text{cm}^{-2}\,\text{s}^{-1} & \text{if } E_n \lesssim 5 \cdot 10^{-2} \ \text{eV} \\ (1.99 \ \pm \ 0.05) \cdot 10^{-6} \ \text{cm}^{-2}\,\text{s}^{-1} & \text{if } 5 \cdot 10^{-2} \ \text{eV} < E_n < 1 \ \text{KeV} \\ (0.75 \ \pm \ 0.09) \cdot 10^{-6} \ \text{cm}^{-2}\,\text{s}^{-1} & \text{if } E_n > 1 \ \text{KeV} \end{cases}$$

$$(5.2.1)$$

where E_n is the neutron kinetic energy. The contribution of the μ-induced neutron component can be evaluated by using the measurements and the extrapolation formula of [AG89b]. With an average muon energy at the MACRO depth $\langle E_\mu \rangle \approx 240$ GeV and a muon flux $\phi_\mu \sim 1$ m^{-2} h^{-1}, the rate of μ-induced neutrons per unitary path length of the muon (in g cm^{-2}) and unitary area is

$$R_{n_\mu} \approx 10^{-11} \left(\frac{\text{n}}{\text{g cm}^{-2}} \right) \ \text{cm}^{-2}\,s^{-1} \tag{5.2.2}$$

To estimate the μ-induced neutron flux, the muon path length must be inserted. A neutron produced in the rock above the detector is re-absorbed before reaching the hall, unless its production point is close to the ceiling. As a rough evaluation we can consider a 5 m path length in the rock, corresponding to ≈ 1300 g cm^{-2}; the μ-induced neutron flux becomes

$$\phi_{n_\mu} \sim 10^{-8} \ \text{cm}^{-2}\,\text{s}^{-1} \tag{5.2.3}$$

The neutron flux produced by the muons crossing the detector would be of the same order of magnitude.

While the low-energy γ's from natural radioactivity are fully rejected by the primary acquisition threshold $E_{th} \sim 7$ MeV, the γ's from neutron capture give an important contribution to the background close to this threshold. Unfortunately, an evaluation of this contribution is extremely difficult, because many different materials can capture neutrons, many decay schemes are involved and the geometry of the apparatus is rather complex. An order of magnitude can be estimated calculating, for instance, how many captures per second R_{capt} can be expected from a horizontal layer of ≈ 100 counters, each one having a 0.6 cm thick PVC ($\rho \approx 1.3$ $g\,cm^{-3}$) box; by taking into account the chemical composition of the PVC and by using the fluxes (5.2.1) and the capture cross section of Chlorine ($\sigma_{Cl} \approx 33$ $barns$ [BN64]), one obtains

$$R_{capt} \sim 4 \ \text{Hz} \tag{5.2.4}$$

The rate of detected de-excitation γ's R_γ is given by R_{capt} times the detection efficiency ϵ of the liquid scintillation counters for γ's; Monte Carlo method calculations give $\epsilon \sim 0.1$ for a 7 MeV energy threshold, so that

$$R_\gamma \sim 0.4 \ \text{Hz} \tag{5.2.5}$$

We expect roughly a $1 \div 2$ Hz contribution from de-excitation γ's at energies above 7 MeV.

5.2.1   Observation of neutron signals in MACRO counters

We searched for neutron signals in MACRO data looking at the distribution of the time difference between consecutive events in the same counter or in adjacent counters.

To do this, we considered the secondary events recorded, after each primary trigger, during the Special Time; the advantage of using secondary triggers is the possibility to study the events over a large energy range ($E \gtrsim 1$ MeV). A neutron is signaled by the de-excitation γ's ($E_{tot} \approx 8$ MeV) following its capture by a material heavier than Hydrogen or by a $\gamma_{2.2}$ following its capture by a proton; therefore we selected two energy regions: $E > 5$ MeV and 1.8 MeV $< E < 2.4$ MeV. We separately studied the secondary events following a *coincidence* and following a *single* (see § (4.6)) primary trigger; here we discuss the second case, where the stronger signal was observed.

First of all we considered the "high-energy" ($E > 5$ MeV) region.

High-energy neutron signal

The energy distribution of the secondary events on the counter horizontal planes is shown in fig. (5.7) between 4 and 10 MeV. This distribution shows a flat plateau on the B and C planes extending from ~ 4 to ~ 6.5 MeV, followed by a decreasing tail up to 9 MeV and beyond; only few events are present on the T plane and their energy distribution is poorer than that of the B, C planes at $E > 5$ MeV.

If one selects the secondary events with energy 5 MeV $< E <$ 10 MeV and looks at the bidimensional distribution of the secondary vs primary longitudinal position, one obtains the plot shown in fig. (5.8). Primary and secondary events are strongly correlated, their longitudinal positions being concentrated along a 45° straight line. The difference between the secondary and primary event reconstructed positions has a quasi-gaussian distribution, with $\sigma_{\Delta x} \approx 140$ cm.

Finally, the distribution of the secondary event delay in respect to the primary events is shown in fig. (5.9); the secondary event energies were selected between 5 and 10 MeV.

In a pure random process, the distribution of the time difference between consecutive events is exponential, with a time constant given by the inverse of the trigger rate. The MACRO trigger rate for $E > 5$ MeV is ~ 100 mHz/counter; the corresponding time constant is ≈ 10 s. Within the Special Time window (800 μs) this distribution is indistinguishable from a flat one. The experimental delay distribution shows a distinct exponential slope, superimposed on the (small) flat background; the fitted exponential time constant is 134 ± 5 μs,

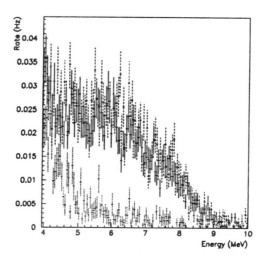

Figure 5.7: Energy distribution of the secondary events in the B (continuous line), C (dashed line) and T (dotted line) planes.

a 25 % lower value than that expected in case of a neutron captured by the liquid scintillator. The content of the first bin is higher than predicted by a pure exponential distribution. In this bin we have contributions from stopping μ decays ($\tau = 2.2 \ \mu s$) and PMT afterpulses, due to traces of gases (mainly Hydrogen and Helium) inside the PMT's; such gases produce peaks at $\approx 0.6 \ \mu s$, $\approx 1.2 \ \mu s$ and $\approx 7 \ \mu s$ after the main pulse [AH91]. A 2 m cut on the distance between secondary and primary events removes the residual background, without significantly affecting the time constant.

We observed ≈ 4000 events of this type in a 2050 hour live time; the corresponding rate is $R_{neu} \sim 50/day$. Note that this rate cannot be compared with (5.2.5) because here only time-correlated neutron signals (i.e. a subsample of all neutron events) are considered.

These events are interpreted as neutrons thermalized by the liquid scintillator and captured by the materials around the detector or (more probably) by the Chlorine in the PVC boxes, after a thermal diffusion inside the liquid. The energy spectrum and the position correlation support this interpretation: the upper portion of the energy distribution extends to the nucleon binding energy, while the lower energy plateau is produced by the only partial containment of the electromagnetic energy corresponding to all γ's in the coun-

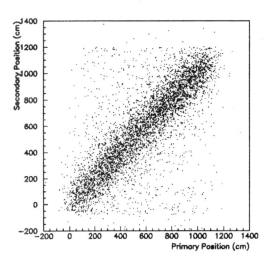

Figure 5.8: Secondary vs primary event longitudinal position; secondary event energies were selected between 5 and 10 MeV.

ters; the difference Δx between the secondary and the primary event position ($\Delta x \lesssim 150$ cm) is reasonable when one considers that the Compton scattering length of a 8 MeV γ (the largest energy expected for the de-excitation photons) inside the liquid scintillator is ~ 70 cm.

To give an explanation of the time constant value we need a model of neutron production and detection. The primary triggers are *single* events; no μ signal is present in the close counters within 1 ms before the neutron event; these observations favour an interpretation of the ^{238}U spontaneous fission as the most important neutron-producing process; therefore we considered a simple model of spontaneous fission events, in which N neutrons and M prompt γ's are emitted.

N has a gaussian distribution; $\bar{N} \sim 3$ for ^{238}U; the total energy released in γ rays is $E_{tot} \sim 6 \div 7$ MeV; the neutrons have a maxwellian energy spectrum: $dN/dE \propto \sqrt{E} \exp(-E/T)$, $T \approx 1.29$ MeV [SE77]. The neutron emission angles are correlated because of kinematical constraints (momentum conservation); for simplicity we first consider an isotropic angular distribution and then discuss the angular correlation effects. The neutron moderation process (which takes $\lesssim 10$ μs and produces a neutron displacement $\lesssim 10$ cm) is neglected; the prompt photon detection is considered instantaneous (if it takes place).

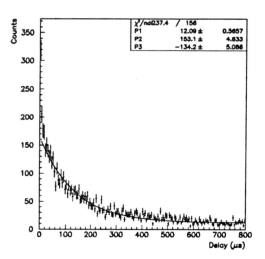

Figure 5.9: Delay distribution between primary and secondary events; secondary event energies were selected between 5 and 10 MeV.

We call t_f the fission event time, t_γ the prompt γ detection time, t_i the ith-neutron capture time and τ the neutron capture time constant; our previous assumption is $t_\gamma = t_f$. The neutrons are ordered in time following their capture sequence. The distribution of all the neutron capture times in respect to t_f is exponential, with time constant τ; the distribution of the delay $\delta_{n,m}$ of the nth neutron in respect to the mth neutron ($n > m$, n and m arbitrary) is given by:

$$\frac{dN}{d\delta_{n,m}} \propto \left[1 - \exp\left(-\frac{\delta_{n,m}}{\tau} \right) \right]^{n-m-1} \exp\left[-(N-n+1)\frac{\delta_{n,m}}{\tau} \right] \qquad (5.2.6)$$

where the first term comes from the probability that $(n-m-1)$ neutrons are captured within the time interval $\delta_{n,m}$ and the second term from the probability that $(N-n)$ neutrons are captured between $t = \delta_{n,m}$ and $t = \infty$ and that the nth neutron is captured between $\delta_{n,m}$ and $\delta_{n,m} + d\delta_{n,m}$. The exponential slope has a time constant $\tau/(N-n+1)$; the distribution (5.2.6) was checked by the Monte Carlo method.

Our signal is a mix of many components, since the following situations are possible:

- the prompt γ's are detected and give the primary trigger; then, the t_n distribution is given by the (5.2.6) with $m = 0$;

- the prompt γ's are not detected; the primary trigger is then given by the first captured neutron whose de-excitation photons release in the liquid scintillation counter an energy deposit larger than the primary threshold. The t_n delay distribution follows the (5.2.6), with m equal to the ordering number of the neutron which gives the primary trigger.

We now discuss the angular correlation effects. For an isotropic angular distribution the capture probability is the same for all the neutrons, since there are no constraints between the neutrons emitted during the same fission event. In a real fission event, the momentum conservation compels the neutron emission angles to be correlated. Consider for instance two neutrons: they are emitted in opposite directions and then, if one enters a counter, the second does not enter it. Neutrons are efficiently moderated by low-A materials and especially by hydrogen-rich substances, like the liquid scintillator; the Gran Sasso rock ($CaCO_3$) has $\langle A \rangle \approx 23$ and is not a good neutron moderator; then, a neutron emitted in the rock is unlikely to be thermalized. The neutron capture probabilities are not independent and only a part of the N neutrons are usually detected.

To study these effects, we modelled the angular correlations introducing a prompt γ detection efficiency and a neutron capture probability which decreases when the ordering number of the neutron increases; an example of the results is shown in fig. (5.10). A pure exponential distribution (top of figure), with a time constant $\tau = 182 \ \mu s$, is distorted (bottom of figure, note the worse χ^2 value) in a roughly exponential distribution, with a reduced time constant: $\tau' = 133 \ \mu s$. The strong similarity of this time constant with that of the observed distribution is artificial, since it is produced by an ad-hoc choice of the efficiencies and capture probabilities; the γ and neutron energy spectra and the neutron moderation are not included. The purpose of figure (5.10) is to show the general effects of combining the (5.2.6) distributions and particularly the time constant reduction.

The neutron source is probably the Gran Sasso rock which fills the steel boxes interspaced between the streamer tube planes; this interpretation is favoured by the observation that such events were practically absent on the *"attico"* counters. If one roughly computes the expected rate of spontaneous fission events produced by the ^{238}U contamination ($\sim 0.5 \times 10^{-6} \ g/g$) of the rock in the boxes, the following value is obtained:

$$R_{spon} \sim 5 \times 10^5 \ fissions/day \tag{5.2.7}$$

When the fission neutron and γ energy spectra, the energy degradation in the rock and the counter detection efficiency for neutrons and γ's are included, this rate is largely reduced and provides a qualitative explanation of the observed events.

We observed also a weaker but qualitatively similar signal after the *coincidence* events. This signal can be interpreted as due to neutrons mainly

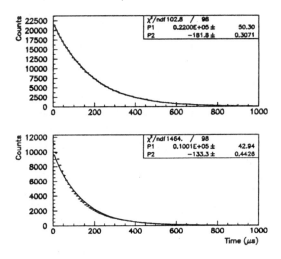

Figure 5.10: Distortion of a pure exponential distribution (top) produced by the insertion of efficiencies and capture probabilities (bottom).

produced by μ spallation processes in the counters or in the rock close to them; the primary triggers are given by the cosmic ray muons.

Low-energy neutron signal

The search for neutron signals in the "low"-energy region (1.8 MeV $<$ $E <$ 2.4 MeV) was much more difficult, since the background rate in this energy window is $\approx$ 200 Hz/*counter*; moreover, the signal is distributed in an acquisition time of $\sim$ 3 months (while, for a supernova explosion, the signal is concentrated in $\lesssim$ 10 s). After applying a position correlation cut ($|x_{sec} - x_{pr}| <$ 1 m), we obtained the delay distribution shown in fig. (5.11). The position correlation cut reduces the background by a factor $\approx$ 6; the background time constant is therefore 1/(200/6) s $\approx$ 30 ms, $\approx$ 40 times larger than the Special Time duration; the background can then be considered approximately flat. The delay distribution was fitted by the sum of a constant and an exponential slope; the time constant obtained by the fit is: $\tau = 181 \pm 35~\mu s$. This time constant would exhibit the same reduction effect observed for the "high-energy" neutrons; the absence of the reduction can be attributed to the larger statistical fluctuations, since the signal is much weaker than in the previous case.

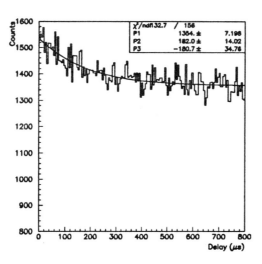

Figure 5.11: Delay distribution between primary and secondary events; secondary event energies were selected between 1.8 and 2.4 MeV and a 1 m position cut was applied.

5.3   Cosmic Ray Muons

The flux of the penetrating cosmic ray muons at the MACRO depth is about $\phi_\mu \sim 1\,m^{-2}\,h^{-1}$, corresponding to a rate $\nu \approx 2.5$ mHz per counter. Muons usually hit two or more counters and are detected by the streamer tube system; the events classified as "coincidences" in § (4.6) are mostly due to cosmic rays. The tracking information is used to reconstruct the muon trajectory and calculate the corresponding energy loss for vertical crossing. The energy loss of the vertical muons in MACRO horizontal counters has an asymmetric Landau-type distribution, with an average $E_{av} \approx 40$ MeV and a most probable value $E_{max} \approx 34$ MeV. Examples of muon energy loss distributions in a horizontal and a vertical counter are shown in fig. (5.12); the energy loss was renormalized to the counter vertical crossing using the streamer tube information.

Muons can be rejected (see later) with an efficiency $> 95\,\%$ by using the timing and position correlation between the scintillation counters and the streamer tube system.

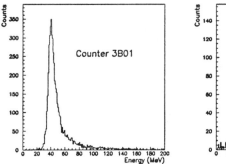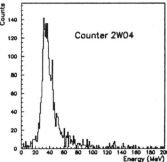

Figure 5.12: Vertical muon energy loss distributions in a horizontal (left) and a vertical (right) MACRO liquid scintillation counter.

5.4   Charged spallation products

The cosmic ray muon interactions with the Gran Sasso rock are a source of neutral (neutrons) and charged spallation products.

A cosmic ray muon can cause a nuclear fragmentation or (if it stops in the apparatus) can be captured by a nucleus; in both cases, radioactive fragments are created. Such fragments de-excite by β and γ emission; their lifetimes range from ms to tens of s and the energies of the β rays can reach ~ 20 MeV. This background is important in searching for very rare events (for instance, solar neutrinos) in experiments having a large continuous active volume, like water Čerenkov detectors [KO92], since it takes place also in the fiducial volume and produces neutrino-like events. The spallation products can be rejected in such experiments by using the time and space correlation with the μ-induced signals.

In liquid scintillator experiments the most important μ spallation reactions are:

a) the μ-induced fragmentation of a ^{12}C nucleus, which can create β and γ radioactive elements;

b) the production, by a μ interaction with a nucleus, of a negative pion; this π^- is then captured by a ^{12}C nucleus;

c) the stopping μ capture by a Carbon nucleus

$$\mu^- + {}^{12}C \rightarrow {}^{12}B + \nu_\mu \qquad (5.4.1)$$

The rate of the a) and b) processes in MACRO can be estimated by the measurements performed by the Kamiokande II collaboration [NA88] for reactions on Oxygen

$$R_{Kam} \sim 20 \ events/day/680 \ tonn \ H_2O \qquad (5.4.2)$$

($E > 7.5$ MeV), where the fiducial volume for searching solar neutrinos is considered. The μ flux and the mean μ energy at the Kamiokande depth are $\phi_{Kam} \sim 4 \ m^{-2} h^{-1}$ and $\langle E^{\mu}_{Kam} \rangle \sim 180$ GeV; if we assume that the probability of these processes scales as $\langle E_\mu \rangle^{0.75}$ (as the μ-induced neutron multiplicity) we obtain:

$$N_{day} \sim 20 \times \left(\frac{600}{680}\right) \times \left(\frac{\phi^{\mu}_{MACRO}}{\phi^{\mu}_{Kam}}\right) \times \left(\frac{E^{\mu}_{MACRO}}{E^{\mu}_{Kam}}\right)^{0.75} \sim 3 \div 5 \ events/day \ (5.4.3)$$

in the 600 $tonn$ of MACRO liquid scintillator. Then, $(1 \div 2) \cdot 10^3$ events are expected in one year from the processes a) and b). Some charged fragments can also be produced within the rock, but since they have an extremely short range, they are re-absorbed in the rock.

The process (5.4.1) is competitive with the stopping muon decay

$$\mu^- \to e^- + \nu_\mu + \bar{\nu}_e \qquad (5.4.4)$$

About 1.4 % of the muons which stop in the scintillator are captured by a Carbon nucleus [MI72]. The ^{12}B nucleus is β-unstable; the end-point of its β spectrum is $E_{max} = 13.4$ MeV for transitions to the ground state and its lifetime $\tau = 20.4$ ms. The number N_y of expected (5.4.1) reactions in MACRO counters in one year can be evaluated as

$$N_y = N^{\mu \ stop} \times f_{^{12}B} \times f_{gs} \times \epsilon \qquad (5.4.5)$$

where $N^{\mu \ stop}$ is the number of stopping muons in one year, $f_{^{12}B} = 0.014$ is the fraction of stopping muons which are captured by a Carbon nucleus, $f_{gs} = 0.97$ [MI72] is the fraction of excited Carbon nuclei which decay to the ground state and ϵ is the detection efficiency. The number of stopping muons in one year is given by:

$$N^{\mu \ stop} = \phi_\mu \times S \times T \times \alpha \qquad (5.4.6)$$

where ϕ_μ is the muon flux at the MACRO depth, S is the MACRO area, T is the measurement time (one year) and α is the muon attenuation factor in crossing the active medium. A further factor 2 must be inserted because only negative muons can be captured by ^{12}C nuclei.

Only the attenuation caused by the liquid must be taken into account, since only a muon stopping in a counter can produce two pulses in spatial and temporal correlation. The MACRO collaboration [MA90] measured the intensity of the cosmic ray radiation as a function of the traversed depth; if the depth-intensity curve is fitted by a pure exponential curve, the attenuation length results $\lambda = (7.57 \pm 0.06) \cdot 10^4 \; g \; cm^{-2}$. The attenuation factor due to the vertical crossing of a counter can then be estimated as

$$\alpha = 1 - \exp\left(-\frac{h\rho}{\lambda}\right) \approx 2.2 \cdot 10^{-4} \qquad (5.4.7)$$

where $h \approx 19$ cm is the liquid level in a horizontal counter and $\rho = 0.87 \; g \; cm^{-3}$ is the liquid scintillator density. Inserting the numerical factors we have

$$N_y = 13\,\epsilon \qquad (5.4.8)$$

Assuming a generous factor 10 for taking into account the μ angular distribution, the number of scintillation counter planes crossed by the muons, the real path in the liquid etc., we obtain ~ 100 (5.4.1) events per year in our apparatus.

The MACRO normal trigger rate (see later, table (5.2)) for "single" events is ≈ 0.75 Hz for $E > 7$ MeV and ≈ 60 mHz for $E > 10$ MeV; the estimated contribution from the processes a), b) and c) is < 0.1 mHz above 7 MeV and is therefore negligible.

A search was performed for these rare events looking at the distribution of the arrival time difference between primary events. The time correlation (if present) would cause an exponential slope to superimpose on the exponential curve due to the normal trigger rate. No statistically significant evidence for a second exponential component was observed in time windows from 1 ms to 20 s.

Figure (5.13) shows an example of this search, the distribution of the arrival time difference between the *single* events ($E > 7$ MeV) within a 600 ms temporal window (with 5 active SM's). The distribution can be well fitted by a single exponential form; the fitted slope (≈ 0.67 Hz) is in agreement with the expectations for a random arrival time distribution.

A double exponential fit to this distribution is unsatisfactory; so, no evidence for this kind of background is present in the data.

5.5   Techniques for background rejection and residual background

Natural radioactivity γ's and γ's from neutron capture produce e^+-like (*"single"*) events in MACRO counters, with an energy spectrum extending up to ~ 10 MeV. The only way to reject (at least partially) such a background is to introduce an appropriate energy threshold. The choice of this threshold is a compromise between a good background rejection and the sensitivity to the largest fraction of a supernova $\bar{\nu}_e$ spectrum. The supernova $\bar{\nu}_e$ energy spectrum (see figs. (1.4), (1.6)) has its maximum at $E \approx 10 \div 12$ MeV; an experiment with a 7 MeV energy threshold is sensitive to a spectral fraction $f \gtrsim 80 \div 85\%$;

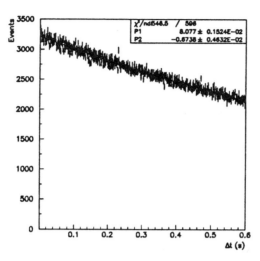

Figure 5.13: Distribution of the delay between primary events ($E > 7$ MeV) for $\Delta t < 600$ ms: no evidence for a second exponential component (superimposed on that due to the normal trigger rate) can be inferred.

we consider this a reasonable fraction; the corresponding background level (see later) is also tolerable. Note that the natural radioactivity background is fully rejected by this threshold; only part of the neutron background survives.

The primary event threshold setting is performed automatically and individually for each counter using the correlation between the threshold and the integral counting rate; a $6 \div 7$ MeV primary energy threshold roughly corresponds to a counting rate $R_{pr} \approx 10$ mHz per counter on horizontal and $R_{pr} \approx 7$ mHz per counter on vertical layers. The counting rate on the whole apparatus is $R_{tot} \approx 3 \div 4$ Hz.

A large statistics energy spectrum of primary events is shown in fig. (5.14) in a semilogarithmic scale; a 7 MeV software energy cut was applied.

The event spectrum is large and steeply decreasing at energies $E \lesssim 10 \div 12$ MeV; it has a deep "valley" between 15 and 25 MeV and a maximum around 40 MeV, with a high-energy tail extending up to 200 MeV and beyond.

The high-energy events can be interpreted as due to cosmic ray muons; the observed spectrum depends on the muon energy spectrum, on the muon crossing direction and on the Landau fluctuations of the muon energy loss. The crossing direction defines the effective path of the muon within the counter, which can be much longer than 19 cm.

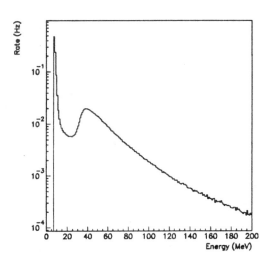

Figure 5.14: Primary event energy spectrum measured in MACRO liquid scintillation counters.

The first group of events can be explained as due to high-energy γ's from
neutron capture and cosmic ray muons going through only a part of a counter.

In fig. (5.15) the event energy and position are shown in a scatter plot E vs
x (with a 7 MeV energy cut); the "low" and "high" energy clusters of events
are clearly visible. Note that the distribution of the event longitudinal position
is flat, as expected, and that only in $\approx 1\%$ of the total number of events the
longitudinal position was erroneously reconstructed outside the counter active
volume.

If all the events seen by two or more counters ("coincidence" events) are
eliminated by using a software selection, one obtains the energy spectrum
shown in fig. (5.16), curve B) (the original spectrum is shown again in fig.
(5.16), curve A), for comparison). This selection has a small effect at energies
$E \lesssim 10$ MeV, but a large one at energies > 12 MeV; the differential rate at
15 MeV is reduced by more than a factor 10 and at 20 MeV by about a factor
20. The quality of the rejection depends on the arrival direction of the muons,
because the scintillation coverage of the apparatus is not perfect; the N and S
faces have no counters at the "attico" level and muons crossing the T layer can
be stopped in the "attico" without traversing the C layer or a vertical layer.

A further reduction can be obtained by rejecting also the events seen by
only one counter and by the streamer tube system within a 5 μs time window,

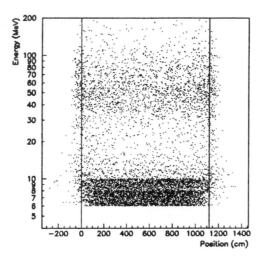

Figure 5.15: Bidimensional distribution of the energy vs longitudinal position for primary events observed in MACRO liquid scintillation counters; the vertical lines represent the limits of the counter active volume.

as in fig. (5.16), curve C); doing so, the inefficiencies of the scintillation counter system and the non-perfect coverage of the apparatus by the scintillators are partially compensated. The distributions B) and C) in fig. (5.16) are similar and differ only by a scale factor at energies > 20 MeV.

The distribution of the residual high-energy ($E > 20$ MeV) events (μ's crossing one counter only) as a function of the number of the counter hit by the muon is shown in fig. (5.17) for the B layer; one observes that these events concentrate near the counters 1 and 16 of each supermodule, close to the apparatus mechanical structure. The interpretation is obvious: the mounting structure, interspaced between the supermodules, breaks the continuity of the active layers.

In fig. (5.18) the integral energy spectra of the events observed in MACRO counters are shown, with and without the "coincidence" event rejection. The "single" event integral rate is considerably reduced, expecially for $E > 20$ MeV, as reported in table (5.2). The values in the fourth column of this table represent the MACRO residual background for e^+-like (primary) events.

The background for $\gamma_{2.2}$-like (secondary) events is due to the natural radioactivity. The integral rate of a counter can be estimated looking at fig.

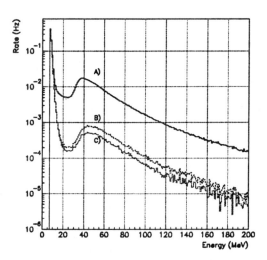

Figure 5.16: Primary event differential energy spectrum measured in liquid scintillation counters; A), B) and C) curves are explained in the text.

(5.4); the integrated counting rate on the "fiducial volume" (the 1.5 m regions close to the PMT's are excluded) is ≈ 700 Hz above 1.5 MeV and ≈ 200 Hz above 2 MeV[‡]. If the full counter length is considered, these numbers are a factor 2 higher. The integral rates with a 1.5 MeV energy cut correspond to about 0.5 secondary events per counter in the fiducial volume and 1 secondary event on the full counter length during each 800 μs secondary time window.

This is a large background, which makes difficult the $\gamma_{2.2}$ detection; however, we already showed (§ (5.2)) that a factor 6 improvement in the signal to noise ratio is obtained by using a 1 m position correlation cut between primary and secondary events. In the next ch. it will be discussed a measurement of the $\gamma_{2.2}$ detection efficiency obtained by means of a γ and neutron $Am - Be$ source. This measurement will be used to evaluate the expected signal to noise ratio for the secondary events in case of stellar collapse $\bar{\nu}_e$'s.

[‡]These values refer to B counters, which have the highest trigger rates; the rates for C, T and vertical counters are a factor 2 lower than those of the B layer.

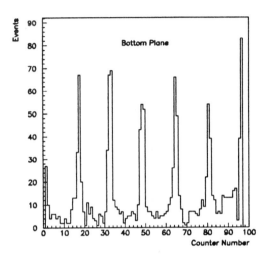

Figure 5.17: Distribution of residual high-energy ($E > 20$ MeV) events as a function of the counter number in the B layer.

Table 5.2: Integrated counting rate in MACRO liquid scintillation counters ($E > E_{cut}$) before and after rejection of "coincidence" events; B) and C) as in fig. (5.16)

Energy cut (MeV)	Rate (Hz) No rejection	Rate (Hz) Rejection B)	Rate (Hz) Rejection C)
7	1.479	0.744	0.731
10	0.773	7.6×10^{-2}	6.3×10^{-2}
20	0.669	3.6×10^{-2}	2.3×10^{-2}
50	0.323	2.1×10^{-2}	1.3×10^{-2}
100	7.0×10^{-2}	4.1×10^{-3}	2.5×10^{-3}

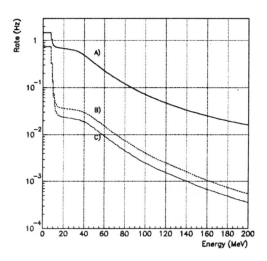

Figure 5.18: Primary event integral counting rate measured in MACRO liquid scintillation counters; A), B) and C) as in fig. (5.16).

Chapter 6

Calibration techniques

Fast and efficient calibration techniques are important "quality factors" for a stellar collapse experiment; a supernova explosion is a very rare event and its correct measurement depends on a reliable and stable detector behaviour and on the knowledge of its energy scale. The good timing and energy resolution of the MACRO counters are fully exploited only if an adequate precision is reached in the determination of the calibration constants. A great effort has been devoted to solve these problems ([BA90b], [BA91a], [BA91d], [BA91e], [BA93a]) and to develop the appropriate calibration techniques.

The goals of the calibration are:

- the definition of a linear (not yet absolute) energy scale by correcting for any non-linearity present in the PMT's, in the front-end electronics, in the *PHRASE* WFD's ...

- the balance between the amplifications of the PMT's mounted at the opposite counter ends;

- the determination of the counter optical attenuation length as a function of the longitudinal coordinate;

- the transformation of the linear energy scale in an absolute one, by the use of a proper energy reference; since the energy range is wide, checks by multiple references are desirable;

- the correction of the time "walk" effect caused by the amplitude variations of the PMT pulses;

- an effective method which keeps to a minimum the time needed for the calibration and for the subsequent data analysis.

6.1   Charge response linearization

The *PHRASE* circuits are optimized for low-energy events; their response is
linear up to ≈ 256 mV input pulses, which correspond to an energy release
(see § (4.4)) $\Delta E \lesssim 15$ MeV at the counter center.

In the linear region, the integration of the digitized waveforms measures the
total light collected by the PMT's and therefore, via (4.5.4), the event energy.
When the pulse amplitude becomes > 256 mV, the Flash ADC saturates and
the digitized pulse becomes apparently wider. This is true for most cosmic ray
muons ($\Delta E \approx 40$ MeV), while for supernova neutrinos the saturation would
be present at the high-energy tail of the spectrum. The area of the digitized
waveforms continues to increase (even if not linearly), as a function of the
energy, so that the event energy can still be determined. To do this correctly, a
linearization algorithm must be developed. (Also the PMT response saturates,
but only at much higher light levels: pulse amplitudes > 5 V).

A linearization method based on and checked by an UV-light laser calibration system has been used; an example of it is shown in fig. (6.1), where
the PMT-*PHRASE* response of a group of counters to increasing intensity
laser light pulses is presented; the ordinate of the curve is the signal integrated waveform, the abscissa is the (arbitrary units) linear laser light intensity.
The non linear response is a monotonically increasing function of the light

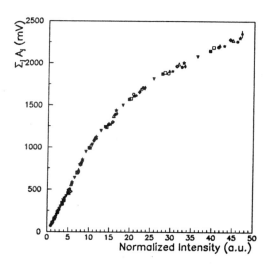

Figure 6.1: PMT-*PHRASE* response to variable intensity laser light.

intensity which can be inverted, providing the value of the "linearized light intensity" corresponding to each measured pulse amplitude. However, fig. (3.6) shows that the UV laser light produces PMT pulse shapes different from the ones of ionizing particles, like cosmic ray muons or γ-ray Compton electrons; the laser-based linearization algorithm must be corrected to include this effect.

To determine the correction, we use the general rule that the response of any physical system to an input signal is given by the convolution of the input signal and of the system transfer function. The *PHRASE* waveform digitizer sampling frequency is 100 MHz; according to the usual prescriptions [PR86], the high frequencies are limited by a 50 MHz input filter, which optimizes the matching between the input pulses and the waveform digitizers.

In the adopted procedure, large samples of waveforms corresponding to cosmic ray muons and to laser light pulses were collected and recorded by a 2 *Gs/s* fast digital oscilloscope (*Tektronix* TDS620) at the *PHRASE* waveform digitizer input. The waveforms were Fourier transformed and bin-by-bin multiplied by the Fourier transform of the *PHRASE* Flash ADC transfer function, the 50 MHz input filter mentioned above; this product gave the Fourier transform of the integrated waveforms at the *PHRASE* output. The Flash ADC integrated waveforms were obtained by Fourier antitransform and approximated by best-fit pulse shapes (a gaussian rise-time function followed by two exponential decay functions). This pulse shape model is used to predict the Flash ADC integrated waveforms corresponding to arbitrary amplitude input pulses of any origin (laser light or ionizing radiation).

A direct check of this Fourier method by natural radiation is impossible (there is no independent measurement of the muon energy loss), but its quality was verified using the laser-light waveform sample. A linearization curve corresponding to the laser light was determined and compared with that directly obtained by the *PHRASE* measurements; the two results were in good agreement. Then, the Fourier method muon linearization function provides the correct extrapolation of the *PHRASE* response to ionizing radiation high energy losses; see fig. (6.2). Note that this linearization function is universal, because all *PHRASE* circuits have the same transfer function, but for a multiplicative amplification constant.

6.2   PMT amplification balancing

PMT amplifications at the two counter ends are in general somewhat different; to correct for these differences ("balancing") we select events close to the counter center, which emit the same amount of light towards both counter ends. Laser light or μ events can be used; the first technique is faster, the second does not require stopping the normal data acquisition.

When using the laser light, the ratio between the integrated waveforms corresponding to the left and to the right PMT is calculated for each laser pulse; the distribution of this ratio is fitted by a gaussian function and the

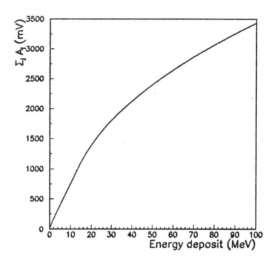

Figure 6.2: Muon linearization function.

mean is used to derive two calibration constants, C_L and C_R, normalized to give a product one; data are collected for several light levels with $20 \div 100$ pulses at each laser setting.

The alternative technique is based on physical events with energies > 10 MeV (mostly cosmic ray muons) and longitudinal position within 1 m from the counter center. For each event, the ratio between the reconstructed left and right "energies" E_L and E_R (see (4.5.3) for the $E_{L,R}$ definition) is computed; the time-averaged value of this ratio is calculated using a $1 \div 2$ week long data-taking period and the calibration constants are extracted as previously explained. In the case of PMT breakdowns or in case of a PMT high-voltage resetting etc., a recalibration is performed limited to the appropriate counters.

6.3   Determination of the counter light attenuation curve

The light attenuation curve is measured by using cosmic ray muons.

Events with energies between 25 and 100 MeV are selected and divided in 40 cm bins of the longitudinal coordinate x ($x = 0$ on the counter left end). The event pulse amplitudes A_L and A_R from the two counter ends are recorded and their average values are computed for each x bin. In the simplest approximation, the pairs (x, A_i) $(i = L, R)$ can be fitted by a pure exponential form; the corresponding attenuation length λ is usually $\gtrsim 12$ m. A better fit is obtained if one uses a phenomenological double exponential form

$$A_L = \alpha_1^L \exp\left(-\frac{x}{\lambda_1^L}\right) + \alpha_2^L \exp\left(-\frac{x}{\lambda_2^L}\right)$$

$$A_R = \alpha_1^R \exp\left[-\left(\frac{L-x}{\lambda_1^R}\right)\right] + \alpha_2^R \exp\left[-\left(\frac{L-x}{\lambda_2^R}\right)\right] \qquad (6.3.1)$$

where λ_1^L, λ_1^R and λ_2^L, λ_2^R are the two attenuation lengths for the left and right side and L is the length of the counter active volume. Typical values of λ_1 and λ_2 are $80 \div 120$ cm and $12 \div 14$ m; those of α_1 and α_2 $70 \div 100$ and $0.5 \div 1$ (in arbitrary units). The two terms in (6.3.1) are similar at $x \sim 2 \div 3$ m (left) or $(L - x) \sim 2 \div 3$ m (right). The expected solid angle dependence ($\propto 1/x^2$, see (3.4.1)) is absorbed in the shorter (λ_1) attenuation length term. In principle a unique attenuation curve would be sufficient for both the left and the right side, but we prefer to determine two independent curves to have more accurate fits.

In fig. (6.3) we show, for the counter $1B01$, the ratios of the pulse amplitudes for events at x and at the counter center ($x = 560$ cm), where the attenuation function value is chosen to be one. The left and right scale are for the left and right pulse amplitudes.

In fig. (6.4) the distribution of the horizontal counter "long" attenuation lengths (on both counter sides) is shown; the mean of the distribution is ≈ 13 m and its width ≈ 3.5 m.

6.4   Determination of the absolute energy scale

Proper energy references are necessary to convert the (arbitrary units) linearized energy scale into an absolute one. Since supernova neutrinos are expected to release in the MACRO counters $\Delta E \gtrsim 10$ MeV, it is desirable to have at least a calibration point in the few MeV energy range; a second higher-energy point is important for checking the linearization procedure. Natural radioactivity provides the former, cosmic ray muons the latter; we already observed that the muon energy deposit in MACRO counters is usually in the *PHRASE* Flash ADC saturation region.

6.4.1   Calibration with natural radioactivity

The 2.614 MeV γ-line emitted by ^{208}Tl (present in the Gran Sasso rock and concrete) is the low-energy calibration point; it was previously stressed that

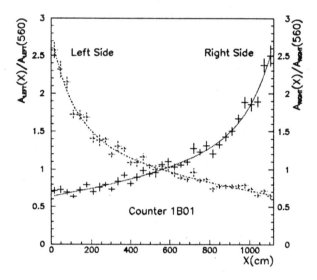

Figure 6.3: Light attenuation curve measured for counter $1B01$. The curves were fitted using (6.3.1).

this line is easily identified by the slope change at the upper end of the radioactivity energy spectra measured by the liquid scintillation counters (see fig. (5.2)).

The usual procedure consists in collecting a large number of secondary events (typically $\sim 1 \div 2 \times 10^4$ per counter, corresponding to about 2 weeks of normal data taking) and in fitting the energy distributions, counter by counter, as in fig. (5.2). The ratio between the nominal and the measured position (mean of the gaussian function) of the ^{208}Tl-line provides the energy scale calibration constant. The Tl-line nominal value is set to 2.37 MeV (with an uncertainty, estimated by the Monte Carlo method, $\sigma_{Tl} \approx 50$ KeV) in order to take into account the energy leakage and the scintillator response saturation, as explained in § (3.7). The analysis time needed is ~ 2 days.

The outlined procedure has the big advantage of using data collected during the normal acquisition, without need for dedicated runs; the disadvantage is that the data collection time is rather long: in a 2 week interval the "natural" instabilities of the apparatus are negligible, but sometimes human interventions can significantly alter the normal working conditions. Moreover, for the vertical counters, 2 weeks of data taking are generally not enough, be-

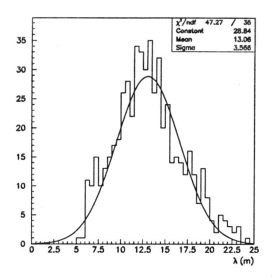

Figure 6.4: Light attenuation length distribution for horizontal counters.

cause of their lower counting rate and poorer energy resolution; an adequate statistics requires ∼ 4 weeks of data-taking.

An alternative procedure uses fast dedicated runs, in which a large number of secondary windows is opened by primary triggers produced by a $0.5 \div 1$ Hz pulser. Two supermodules can be calibrated at the same time; their normal acquisition must be obviously stopped. One-day data taking is needed to collect ∼ 50000 secondary triggers for all counters of the SM's under calibration; then the energy spectra are fitted as in fig. (5.2). The advantage of this second procedure is its speed, the disadvantage is the need for a partial modification of the data acquisition; this method is therefore less frequently used. The dedicated run technique is particularly advisable when a general PMT gain resetting is performed; a total time of 4 days is required to collect and process the data and determine a complete set of calibration constants for the whole apparatus.

The precision obtainable with the Tl-line method was evaluated by using a more complex analysis chain, based on the assumption that all counters belonging to the same layer have similar differential energy spectra (far from the threshold), provided that all calibration constants were correctly inserted. A check of this hypothesis is shown in fig. (5.3). Suppose that one calibrates a counter energy spectrum with very good accuracy; this spectrum can be used as a reference and the spectra of all the other counters belonging to the same layer can be compared to it. The calibration constant of each counter must be chosen to make its energy spectrum as similar as possible to the reference one; this is automatically done by using a χ^2-minimization algorithm.

Let α_j be the calibration constant needed to correct the energy scale of the jth counter and consider an energy interval Γ_j reasonably far from the energy thresholds of both the jth and the reference counter. Divide Γ_j in N_j bins and consider the quantity:

$$\chi^2(\alpha_j) = \sum_{i=1}^{N_j} \left\{ \frac{n_i^{ref} - n_i^j(\alpha_j)}{\sqrt{n_i^j(\alpha_j)}} \right\}^2 \qquad (6.4.1)$$

where n_i^{ref} is the number of counts in the ith bin of the reference counter spectrum and n_i^j is the number of counts in the ith bin of the jth counter spectrum. n_i^j depends on α_j because the content of each bin is determined by the energy scale. The α_j value is automatically computed by an iterative procedure, which selects the Γ_j interval and searches for the minimum of the $\chi^2(\alpha_j)$ function. The α_j constants are inserted to scale all the spectra and superimpose them to the reference spectrum; the positions of the Tl-lines in the rescaled spectra are then determined by the usual fitting procedure. The precision in determining the Tl-line position is extracted looking at the distribution of the rescaled ^{208}Tl peaks relative to the reference spectrum peak (set at 2.37 MeV). This distribution for a sample of 70 horizontal counters is shown in fig. (6.5); it is well fitted by a gaussian shape. The mean value of the gaussian curve is in good agreement with the nominal position of the Tl-line and its relative width ($\sigma/\bar{E} \approx 3\%$) is a measurement of the accuracy of the technique.

6.4.2   Calibration with cosmic ray muons

Cosmic ray muons provide a "high"-energy calibration point, which can be used as an alternative to the Tl-line or as a check of the linearization and energy reconstruction procedure.

The muon energy loss is renormalized to a vertical crossing by using the streamer tube information; the energy calibration constant of each counter is extracted by comparing the measured muon energy loss, at the Landau peak, with its expected value, which can be calculated from the height of the counter liquid layer. In fig. (6.6) the vertical muon energy loss in a horizontal MACRO counter (1B04) is shown and compared with the Monte Carlo method calculation of the expected Landau distribution for relativistic particles vertically crossing the liquid scintillator; the simulated muon direction is extracted from the experimental angular distribution (measured by the streamer tube system); the energy loss is calculated using the path in the liquid scintillator and it is

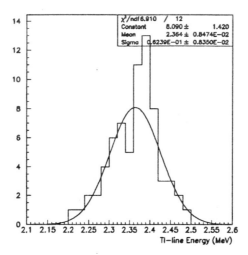

Figure 6.5: Accuracy of ^{208}Tl calibration technique for 70 horizontal counters. The ratio between the width and the mean of the gaussian fit is a measurement of the accuracy: $\sigma/\bar{E} \approx 3\,\%$.

renormalized to a 19 cm vertical crossing. Note that the Landau tail is correctly reproduced. The width of the distribution can be measured fitting by a half-gaussian form the portion of the spectrum on the left of the Landau peak; the typical values are $\sigma_H \sim 3 \div 4$ MeV for horizontal and $\sigma_V \sim 5 \div 7$ MeV for vertical counters.

Usually some thousand μ's represent a sufficient statistics; 2 weeks of data correspond to $\sim 2500 \div 3000$ μ's for each horizontal counter and to $\sim 1500 \div 2000$ μ's for each vertical counter. A few per cent accuracy on the Landau peak position can be obtained for all counters.

In fig. (6.7) is shown the distribution of the ratio between the renormalized Landau peak and Tl-line position for 287 (over a total of 294) horizontal counters. The mean ratio between the μ and Tl positions (determined by a gaussian fit) is 14.9, in agreement with the nominal value (≈ 15) expected from the energy loss calculations; the sigma of the distribution is $\sigma_H \approx 0.9$ and checks the accuracy of our hypothesis of the universality of the energy scale linearization function. (For vertical counters the mean value is $R_V \approx 14$, in reasonable agreement with expectations, but $\sigma_V \approx 4$). We assume that, for each counter, the measured position of the μ peak and of the Tl-line are in

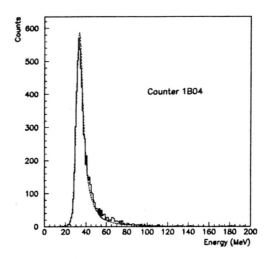

Figure 6.6: Vertical muon energy loss in a horizontal MACRO counter; the superimposed curve is the expected Landau distribution for relativistic particles.

agreement if their ratio is approximately within $3\,\sigma_H$ from the nominal value, i.e. if $12.5 < R < 17.5$; this "$3\,\sigma$" tolerance accounts for the possible differences among the *PHRASE* circuits.

6.5 Measurement of the neutron detection efficiency

The neutron detection efficiency in MACRO counters was measured by a low intensity (emission rate $R \approx 2700\,n\ \mathrm{s^{-1}}/4\,\pi$) uncollimated *Am/Be* source [BA91a]. Neutrons are produced by the reactions shown in fig. (6.8); 60 % of the neutrons are associated with a cascade de-excitation photon from the 4.438 MeV $^{12}C^*$ level (from now on $\gamma_{4.4}$). The *Am/Be* source was placed outside the counters, at the center of their longitudinal face, so that the source longitudinal position was ≈ 560 cm. No other device was needed in addition to the liquid scintillation counters and the associated *PHRASE* circuits.

Primary and secondary *PHRASE* thresholds were set at $E_{pr} \approx 3$ MeV and $E_{sec} \approx 1$ MeV; this unusually low value of E_{pr} allowed the detection of $\gamma_{4.4}$ as primary events. When a neutron, emitted in association with a $\gamma_{4.4}$ primary,

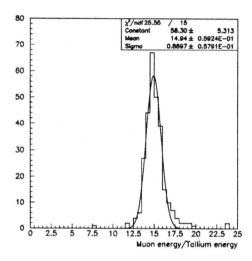

Figure 6.7: Distribution of the ratio between the normalized μ peaks and the ^{208}Tl line positions for 287 horizontal counters.

is thermalized and captured in the liquid, the $\gamma_{2.2}$ secondary event is time correlated with the $\gamma_{4.4}$.

During this measurement a large radioactivity background (rate $\sim$ 5 kHz) was present; we measured the background spectrum by opening $\sim$ 300000 secondary windows with a high-frequency pulser; the signal + background spectrum, obtained when the secondary window is opened by a $\gamma_{4.4}$, and the pure background spectrum were normalized at the same number of primary triggers and the renormalized background spectrum was subtracted from the signal + background spectrum; the result is shown in fig. (6.9a) (upper part). The spectrum is well reproduced by a Monte Carlo method simulation [BA91c]. The lower part of this figure shows the decomposition of the observed signal into its main contributions: a) $\gamma_{4.4}$ accidentally detected while the energy threshold is low; b) correlated and uncorrelated $\gamma_{2.2}$ (the latter come from neutrons not associated with the $\gamma_{4.4}$ primary) and c) recoil protons produced by neutron-proton scattering in the liquid. The number of detected events agrees with the absolute Monte Carlo method predictions (no relative normalization is needed) within 10 % for both $\gamma_{4.4}$ and $\gamma_{2.2}$.

The distribution of the secondary event reconstructed positions along the counter (1 MeV $< E_{sec} <$ 3 MeV) is shown in fig. (6.10); the width of the peak in this figure provides an upper limit (the source was uncollimated) for

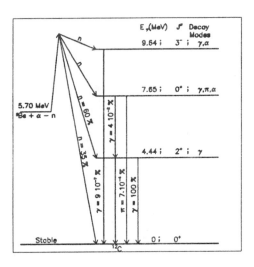

Figure 6.8: Energy level diagram showing electromagnetic transitions to the ^{12}C ground level following a $^{9}Be\,(\alpha,\,n)\,^{12}C^{*}$ reaction.

the longitudinal position resolution: $\sigma_x < 1$ m. This resolution is worse than that of the primary reaction (2.1.1): $\sigma_x^{e^+} \approx 25$ cm; a 1 m-spatial correlation cut is the selection criterion we adopt to identify correlated $\gamma_{2.2}$ events.

The distribution of the delay between primary and secondary triggers is shown in fig. (6.9b) for 1 MeV $< E_{sec} < 3$ MeV and a position cut $|x_{pr} - x_{sec}| < 1$ m. The distribution consists of a decay exponential function superimposed on a flat background. The background is due to the uncorrelated components of the source induced signal discussed above. The characteristic delay time is $\tau \approx 180~\mu s$, as expected from neutron capture in the liquid scintillator. The efficiency for detecting the neutron capture following a primary $\bar{\nu}_e$ event, estimated from these data, is $\epsilon_{2.2} \approx 25\,\%$ for a secondary software energy threshold $E_{sec} \approx 1.5$ MeV.

The compromise between signal and noise on secondary events is much harder than on primaries, because the radioactive background spectrum is a very steep function of the energy. Reducing the energy threshold from 1.5 MeV to 1 MeV causes an increase in $\epsilon_{2.2}$ from $\approx 25\,\%$ to $\approx 30\,\%$, but the corresponding increase of the background integral rate is a factor ≈ 2; on the other hand, a 2 MeV secondary energy threshold reduces the background rate by a factor ≈ 4, but $\epsilon_{2.2}$ drops to $\approx 18\,\%$. The 1 m-spatial correlation cut is a basic tool for selecting the neutron signal, since it increases the expected

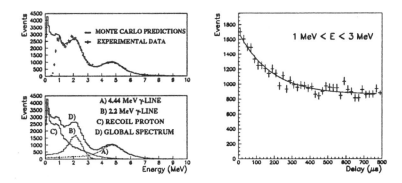

Figure 6.9: (a) Observed and predicted energy spectrum for secondary events induced by a *Am/Be* source. (b) Distribution of the delay between primary and secondary events for 1 MeV < E_{sec} < 3 MeV (position cuts were applied).

signal to noise ratio, for a 1.5 MeV energy threshold, from $\approx$ 0.27 to $\approx$ 1.5 in the full counter length and from $\approx$ 0.4 to $\approx$ 2.3 in the fiducial volume (the regions corresponding to the last 1.5 m close to the PMT's are excluded). A further improvement can be obtained by using a narrower (400 μs) time window, which reduces the background by a factor 2 at the expense of a $\approx$ 10 % signal loss. Optimizing all these cuts, we obtain a $\epsilon_{2.2} \approx 22$ % efficiency on the $\gamma_{2.2}$ detection and a signal to noise ratio $\approx$ 2.7 in the full counter length and $\approx$ 4 in the fiducial volume for a 1.5 MeV software energy threshold.

6.6   Leading edge corrections

The event longitudinal position reconstruction is based, as already stated, on the difference between the light transit times towards the two counter ends (see (4.5.1)); the event arrival time at one counter end is defined as the time at which the PMT signal crosses the corresponding *PHRASE* discriminator threshold.

Events of different energy (at a fixed position) produce, on the same PMT, pulses of different amplitude, which are subjected to the time "walk" effect [LE92] (fig. (6.11)). Suppose that two coincident signals with different amplitudes are introduced in a discriminator; the larger pulse will trigger the discriminator before the smaller pulse; note that the pulses are equal in width and shape, the unique difference being their amplitude.

To be more quantitative, consider a pulse having a half-gaussian leading edge; if t^* is the time corresponding to the pulse maximum, the time t when

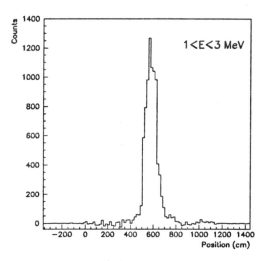

Figure 6.10: Distribution of the secondary event reconstructed position along the counter. The width of the peak is an upper limit for the longitudinal position resolution. The source position is at the counter center (560 cm).

the pulse crosses the discriminator threshold $\bar{V}$ is the root of the equation

$$A \exp\left\{-0.5\left(\frac{t - t^*}{\sigma}\right)^2\right\} \Theta\left(t^* - t\right) = \bar{V} \qquad (6.6.1)$$

where A and σ are the pulse amplitude and width and the $\Theta\left(t^* - t\right)$ function ensures that only the solution at $t < t^*$ is retained. t^* can be interpreted as the time when a pulse, having an amplitude equal to the threshold $(A = \bar{V})$, triggers the discriminator. The solution of (6.6.1) is

$$t = t^* - \sqrt{2}\,\sigma\sqrt{\ln\left(\frac{A}{\bar{V}}\right)} \qquad \left(A > \bar{V}\right) \qquad (6.6.2)$$

The time "walk" effect can be measured (and software corrected) by using the variable intensity laser based calibration system.

A high intensity laser light beam is sent towards a reference PMT; the same beam is sent, through a variable attenuator, towards the counters and the difference between the signal arrival time at each counter end and at the reference PMT is measured. The variable attenuator is used to produce laser

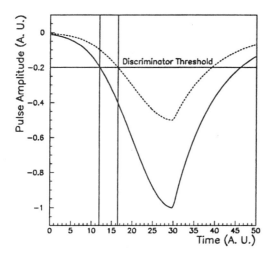

Figure 6.11: The "walk" effect: two coincident signals cross the discriminator threshold at two different times, because of their difference in amplitude.

light pulses of different intensities, i.e. different amplitudes; the relation between the pulse amplitude and the signal arrival time difference can then be explored. An example of these measurements for the counter $6B01$ is shown in fig. (6.12); the intensity of the laser light is given in units of equivalent energy at the counter center. The pairs (energy, delay) are fitted by the (6.6.2), with $\bar{V}$, σ and t^* as free parameters ($\bar{V}$ is also given in units of equivalent energy); this fit is shown superimposed on fig. (6.12). The typical values of the parameters are $\sigma \sim 8$ ns, $t^* \sim 50$ ns and $\bar{V} \sim 1$ MeV.

The time delay is composed by a common offset ($\Delta t \approx 25$ ns) due to the experimental setup used for the measurement (cable length, electronic chain etc.) and by an amplitude-dependent term, which reaches ~ 30 ns for very low amplitude pulses (equivalent energy $E_{eq} \sim 1$ MeV). A 30 ns delay corresponds to a position difference $\Delta x \sim 6$ m, the speed of light within the counters being $v \approx 20$ cm/ns.

The effects of the leading edge corrections on the longitudinal position resolution can be observed in fig. (6.13), where a B-plane low-energy ($E < 2$ MeV) event longitudinal position distribution is shown without (continuous line) and with (dashed line) the application of the leading edge corrections. Note the significant improvement near to the counter ends. The total number of events below the 2 MeV energy cut decreases when the leading edge corrections are

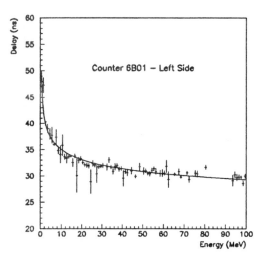

Figure 6.12: Measurement of the pulse delay of the PMT's on the counter $6B01$ (left end) compared with the reference one.

applied, since the energy is underestimated if the event position is reconstructed outside the counter active volume. The use of the leading edge corrections reduces the fraction of badly reconstructed events and makes the energy reconstruction more reliable at very low energies ($E \lesssim 2$ MeV).

The effect of the leading edge corrections on the longitudinal position distribution of the high-energy (e.g. cosmic ray muons) events is negligible.

6.7   Fast check of calibration constants

An original and fast method for determining and checking the calibration constants has been recently developed [BA93a]; it is based on the single-PMT counting rates and can be used in a parasitic way during the normal data acquisition.

This method is based on the hypothesis that the natural radioactivity γ's are uniformly distributed in the apparatus or, more precisely, in each counter layer. We already showed (see ch. 5) that the events generated by the natural radioactivity background, when reconstructed in position and energy, have a rather uniform distribution along the scintillation counters; moreover, the energy spectra measured by the counters belonging to the same layer are

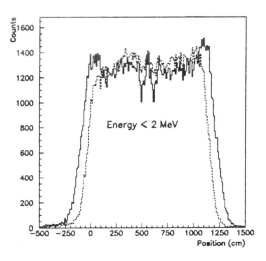

Figure 6.13: Low-energy ($E < 2$ MeV) event longitudinal position distribution without (continuous line) and with (dashed line) the application of the leading edge corrections.

similar, when the proper calibration constants are used. The light produced by the natural radioactivity has a well known distribution and a well defined rate; when this light is observed by the PMT's mounted at one of the two counter ends, one obtains a pulse height distribution given by the convolution of the "source" properties and of the counter light attenuation characteristics. If the scintillation counters are similar and if the hypothesis of uniform radioactivity distribution is correct, the PMT pulse height spectra should be universal in character; a unique function would be adequate for fitting all PMT distributions.

The PMT counting rates are recorded using a simple CAMAC acquisition system, consisting of 6×16-channel programmable threshold discriminators and 3×32-channel scalers produced by PHILLIPS and a Macintosh computer. The discriminator thresholds can be varied in steps of 2 mV. The PMT signals are taken directly from a free output of the MACRO FAN-OUT's and fed to the discriminators by 8-pin flat AMP cables; up to 96 signals (one for each counter end, so that 96 signals correspond to 3 horizontal counter layers) can be studied at a time. The acquisition system is based on the *LABVIEW* [LA94] software; fixed and variable threshold measurements are possible; different acquisition times can be programmed for each threshold value.

The single-PMT counting rate integral spectrum is a rapidly decreasing function of the pulse amplitude. An equal behaviour is expected if one plots the integral counting rate versus the acquisition threshold; varying this from 20 to 300 mV, the integral rate decreases from $\approx 10^4$ Hz to ≈ 10 Hz. The spectrum is well fitted by the empirical form

$$R = \exp\left\{-\left(\alpha_0 + \alpha_1 T + \alpha_2 T^2 + \alpha_3 T^3\right)\right\} \qquad (6.7.1)$$

where R is the rate and T the acquisition threshold. All the spectra of the horizontal and all the spectra of the vertical counters are rather similar and form two distinct groups; each spectrum can be fitted by (6.7.1) and the spectra of each group can be superimposed by a simple change of the T-axis scale by appropriate multiplicative factors; these factors can be identified as the (relative) amplifications of the PMT + FAN-OUT channels. Some examples of PMT curves are shown in fig. (6.14), before (continuous line and circles) and after (triangles) the application of the re-scaling factors.

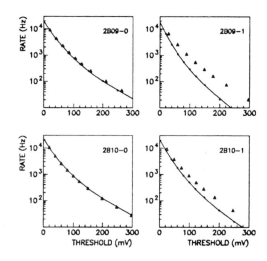

Figure 6.14: Counting rate integral spectra of 2 horizontal counters (left ("0") and right ("1") side) as a function of the acquisition threshold before and after re-scaling.

The multiplicative factors are determined by a χ^2-minimization procedure;

the superimposed spectra of 16 horizontal counters of SM 2 (left side) are shown in fig. (6.15).

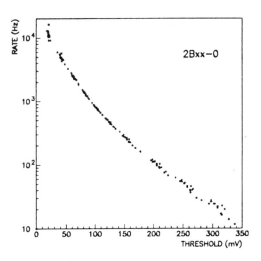

Figure 6.15: The superimposed spectra of 16 horizontal counters of SM 2 (left side).

The characteristics of the PMT counting rate spectrum can be reproduced by a simple model, which uses the measurements of the natural radioactivity by NaI and Ge detectors ([AR92a], [AR92b]) and of the counter light attenuation curve (see § (6.3)). The high-resolution NaI and Ge spectra were re-binned and approximated by a three-exponential function for modelling the effects of the scintillation counter poorer energy resolution. In fig. (6.16) we show the comparison between the SM 2 horizontal counter single PMT spectra (dots) and the simulated one (histogram); the agreement is rather good.

A simple application of the counting rate method is the determination of the PMT amplification balancing factors. The use of laser pulses and cosmic ray muons to determine these factors was previously discussed; the counting rate measurement provides an alternative technique.

The amplification factors can be inferred from the integral spectra by observing that equal amplifications should correspond to equal integral counting rates, if these are measured at the same threshold. One chooses therefore a reference PMT (normally the one with the higher statistics) and compares the integral spectra of the other PMT's with that of the reference; such compar-

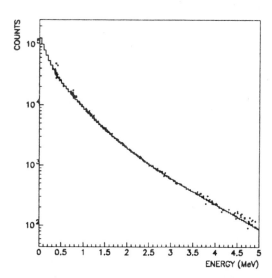

Figure 6.16: The simulated counting rate integral spectrum (histogram) compared with the experimental data (dots).

ison is automatically performed by the χ^2 minimization procedure described above. The multiplicative constants which rescale the spectra to produce the best agreement with the reference one are the PMT + FAN-OUT (relative) amplification factors; then, for each counter, we have a pair of constants, which, after a renormalization to give a product one, are the correct balancing factors for the counter under study.

These factors can be compared with that obtained by the laser technique defining an "asymmetry" variable:

$$\eta = \frac{2\,(A_0 - A_1)}{(A_0 + A_1)} \tag{6.7.2}$$

where A_0 and A_1 are the pulse amplitudes on the left (0) and right (1) PMT. The plot of the Rate asymmetry η_R against the Laser asymmetry η_L for 120 counters is shown in fig. (6.17). The two asymmetries lie along a 45° straight line, i.e. they are strongly correlated and the two methods are in good agreement.

The η_L and η_R variables can not be used to indipendently estimate the precision of the two methods, σ_L and σ_R, since their distributions are deter-

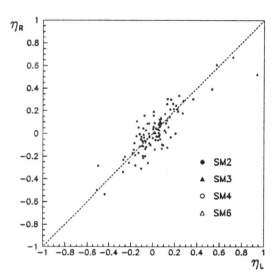

Figure 6.17: Scintillation counter balancing: comparison between the Rate asymmetry η_R and the Laser asymmetry η_L for 120 counters.

mined by σ_L, σ_R and the dispersion of η, i.e. the fact that different counters have in general different asymmetries. An upper limit on the precision of the two methods can be set introducing two new coordinates, $\eta_+ = \eta_L + \eta_R$ and $\eta_- = \eta_L - \eta_R$. By using the difference we remove the correlation between η_L and η_R and the width of the η_- distribution is determined by the precision of the two methods only. The η_- variable has a gaussian distribution with $\sigma_{\eta_-} = \sqrt{\sigma_L^2 + \sigma_R^2} \approx 9\,\%$; then, both the methods have a precision better than $9\,\%$. The same comparison was performed with the cosmic ray muon method, obtaining similar results.

In the natural radioactivity energy range all the electronic chain (PMT's, FAN-OUT's, Flash ADC's etc.) behaves linearly; the PMT amplification factors define therefore a (relative) scintillation counter energy scale. The definition of an absolute energy scale requires an energy reference, like the ^{208}Tl-line. One can calibrate the energy scale of the reference counter, determining a common multiplicative factor, and then transform the relative scales of all other counters into absolute scales simply by introducing this common amplification factor.

The calibration constants obtained by the rate technique can be compared with those deduced by the Tl method; the comparison is shown in fig. (6.18) for a sample of 75 horizontal counters (a scintillation counter corresponds to two experimental points, one for each counter end). The two sets of calibration

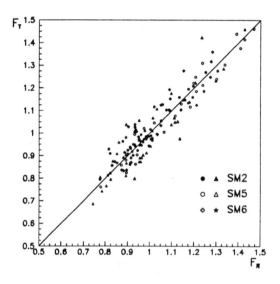

Figure 6.18: Scintillation counter calibration factors: comparison between the rate method factors F_R and the Tl method factors F_T for a sample of 75 counters.

constants are strongly correlated.

The accuracy of the counting rate method can be studied in a way similar to that previously adopted for the Tl method. One measures the counting rate spectrum S of the reference counter and the S_i spectra of the other counters; then the spectra are superimposed, the calibration constants α_i are determined and the spectra S_i and their fit functions F_i are re-scaled by α_i. At this point, a threshold value T is selected (for instance 100 mV) on the reference spectrum S and the corresponding rate $R = R(T)$ is singled out; then one looks, on the re-scaled S_i spectra, for the value T_i corresponding to the same rate R. The distribution of the variable $x_i = (T_i - T)/T$ is shown in fig. (6.19) for horizontal and vertical counters; the widths σ^R of the two distributions are: $\sigma_R^H = (2.5 \pm 0.3)$ % for the horizontal and $\sigma_R^V = (4.6 \pm 0.5)$ % for the vertical counters. These widths are largely independent from the selected threshold value T and represent the accuracy of the counting rate method.

The counting rate method is a simple tool for determining the MACRO counter calibration constants; it is completely parasiting, does not require a

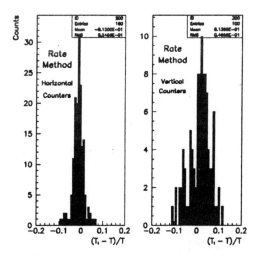

Figure 6.19: Accuracy of the counting rate method for horizontal (left) and vertical (right) counters.

complex analysis chain and its statistical accuracy is at least comparable with that of more refined methods. However, cosmic ray and Tl-line techniques are probably safer, since they are based on energy "lines" and the full event-by-event energy reconstruction; they also take into account all the differences due to the *PHRASE* circuits. The counting rate method, being faster and compatible with the normal data acquisition, is frequently used to determine the calibration constants and the PMT gains and to check their stability, while the Tl-line and cosmic ray methods are applied at longer time intervals (about every 2 months) to obtain a more controlled and reliable definition of the energy scale.

Chapter 7

Results of the search for stellar gravitational collapse neutrinos

7.1 MACRO sensitivity to neutrinos from a stellar collapse

The sensitivity of an experiment to neutrinos from stellar gravitational collapses depends on the detector mass and on the background level. Suppose that $\bar{N}$ is the number of expected events (signal + noise) in a detector for a stellar collapse at a given distance; the sensitivity of the experiment is limited by the probability that a noise statistical fluctuation produces a cluster of $N \geq \bar{N}$ events. Consider one event recorded at $t = t_0$; if R is the experimental background rate, the probability of observing $N - 1$ events in a time interval τ starting at t_0 is given by the Poisson formula:

$$P_N(R, \tau) = \frac{(R\tau)^{(N-1)}}{(N-1)!} \exp(-R\tau) \qquad (7.1.1)$$

where the probability has been labeled "N" instead of "(N - 1)" because N is the total number of events associated with the interval τ. The probability that a background fluctuation simulates a cluster of at least $\bar{N}$ events is

$$
\begin{aligned}
P_{N \geq \bar{N}} &= \sum_{j=\bar{N}}^{\infty} \frac{(R\tau)^{(j-1)}}{(j-1)!} \exp(-R\tau) = \\
&= 1 - \sum_{j=1}^{\bar{N}-1} \frac{(R\tau)^{(j-1)}}{(j-1)!} \exp(-R\tau) \qquad (7.1.2)
\end{aligned}
$$

The average number of events recorded by the detector is given by RT, where T is the experiment working time. A time interval τ is opened after each event; then, a cluster of $\bar{N}$ events can be simulated in a total of RT "attempts". If a new event occurs during the time interval τ, a new gate is opened; this new interval is correlated with the previous, since at least one event is common to the two intervals. The probability of a fake cluster in one interval is then related to that on the previous intervals. If the intervals were uncorrelated, the fake cluster probability in each of them would be $P_{N \geq \bar{N}}$; let P_c be the probability that the intervals are correlated; the larger is P_c, the larger is the difference between $P_{N \geq \bar{N}}$ and the true probability P_t of a fake cluster. P_c is the probability that at least two events (including that at the gate beginning) occur within τ, or $P_c = P_{N \geq 2} = 1 - \exp(-R\tau)$; then, $P_c \ll 1$ if $R\tau \ll 1$. In this case $P_t \approx P_{N \geq \bar{N}}$ and the probability of at least one fake cluster within the measurement time T is well approximated by the binomial formula:

$$
\begin{aligned}
\mathcal{P}\left((\mathcal{R}\tau), \mathcal{T}\right) &= B\left(n > 0, (RT), P_{N \geq \bar{N}}\right) = \\
&= 1 - B\left(0, (RT), P_{N \geq \bar{N}}\right) = \\
&= 1 - \left(1 - P_{N \geq \bar{N}}\right)^{RT}
\end{aligned}
\tag{7.1.3}
$$

where $B(k, n, p)$ is the binomial distribution for a subset of k "successful attempts" in a set of n, each one having a success probability p

$$
B(k, n, p) = \binom{n}{k} p^k (1 - p)^{(n-k)}
\tag{7.1.4}
$$

The MACRO sensitivity to a stellar gravitational collapse is presented in fig. (7.1), which shows the allowable background rate corresponding to three fixed probabilities (7.1.3) for observing a false supernova signal in 10 years for a specified number of primary events; the burst duration is kept fixed at 2 s. The use of the fig. (7.1) can be explained by an example. Suppose that the expected number of events in the τ interval is 10; one can be interested in knowing what is the maximum tolerable background rate if a probability $P \leq 10^{-1}$ for observing a false supernova signal is required. One can draw a $N = 10$ line parallel to the abscissa up to intersect the $P = 10^{-1}$ probability curve; the abscissa of the intersection point is the corresponding allowable rate: $R \approx 100$ mHz. On the other hand, at a fixed rate one can be interested in knowing what must be the minimum number of events in a cluster to have a probability $P \leq 10^{-1}$. In this case, one can draw a line parallel to the ordinate up to intersect the probability curve; the ordinate of the intersection point is the corresponding number of events.

In a search for supernovæ the signal to noise ratio has to be optimized, in order to keep at a minimum the fake cluster probability. Increasing the time interval beyond a few seconds, will capture most of the signal, at the expense of an increased background; on the other hand, reducing the time interval will reduce the background, but the number of expected events will be correspondingly lower. Following numerical simulations and *SN1987A* data, a fraction $\approx 70 \div 75\%$ of the total supernova signal is expected within a 2 s interval. The tradeoff between signal and background is largely determined by the analysis energy cut, which minimizes the noise, but reduces the signal detectable

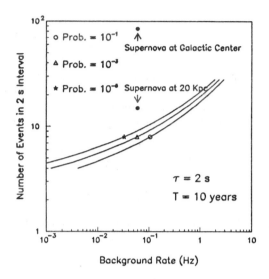

Figure 7.1: Probabilities for observing a given number of events in a 2 s interval for a detector running time of 10 *years* as a function of the background rate.

fraction. Monte Carlo method simulations predict a signal detectable fraction $f_{10} \approx 65\%$ for $E_{th} = 10$ MeV and $f_7 \approx 82\%$ for $E_{th} = 7$ MeV. Assuming a (conservative) value of 150 events expected in MACRO from a supernova at the Galactic Center for a 7 MeV threshold, the number of events which will be recorded within a 2 s interval for a 10 MeV threshold can be computed

$$N_{ev} \approx \begin{cases} 85 \div 90 & \text{for } R_{SN} = 8.5\,\text{Kpc} \\ 15 \div 16 & \text{for } R_{SN} = 20\,\text{Kpc} \end{cases} \qquad (7.1.5)$$

where R_{SN} is the supernova distance from the Sun (these representative distances were chosen taking into account the Galaxy model of fig. (1.11)). The MACRO background rate above 10 MeV is $R \approx 60$ mHz (see ch. 5); the corresponding number of events in 2 s is 0.12 and $P_c \approx 0.11$. The condition $R\tau \ll 1$ is reasonably satisfied and the formula (7.1.3) is adequate for our particular case. Once a supernova candidate is found, the neutrino burst can be studied on a larger time window.

The number of primary events includes the background, but this contribution is practically negligible, because one expects < 0.2 background events in 2 s and < 1 background event in 10 s. The estimations (7.1.5) are indicated by arrows in figure (7.1); the probability of a false supernova signal in 10 *years* is $\ll 10^{-5}$ for a gravitational collapse anywhere in the Galaxy. For a stellar collapse in the Large Magellanic Cloud ($d \approx 50$ Kpc), $\approx 4 \div 5$ neutrino events are expected; such a burst could not be accepted as a supernova on the basis of probability arguments alone, but it could give an independent confirmation of a stellar collapse signal reported by a more massive detector[§] or by optical observatories.

Note that the probability is a smooth function of the background rate; for a 10 mHz rate, a false cluster probability $P \approx 10^{-5}$ in 10 *years* is achieved requiring 6 positrons and for a 100 mHz rate requiring 11 positrons in a 2 s interval. If we write for the probability $\mathcal{P}$ the approximate form

$$\mathcal{P} = C \frac{(R\tau)^\alpha}{N^\beta} \qquad (7.1.6)$$

(R and N are the background rate and the number of expected events and C is a constant) we have from the previous observations: $\alpha < \beta$ (at a fixed τ the same probability is obtained doubling the number of events and multiplying the rate by a factor 10). In a rate interval suitable for MACRO ($10 \div 100$ mHz) $\alpha \approx 4$ and $\beta \approx 13$. The number of expected events N increases linearly with the detector mass M, while the behaviour of the background rate R as a function of M depends on the apparatus geometry; in self-shielded experiments, like *LVD*, R increases less than linearly with M, while in MACRO a linear relationship can be assumed. Choosing, to be conservative, a linear relation for both R and N and inserting it in (7.1.6), we obtain that the probability is a decreasing function of the detector mass M, i.e. the experimental sensitivity is enhanced when the mass is increased.

7.2   The "Early Watch" system for a fast supernova identification

The *SN1987A* optical flare was recorded about three hours after the Kamiokande II - IMB3 neutrino burst. This time delay is in agreement with the calculations of the propagation time of the core collapse shock-wave to the stellar surface and of the supernova optical brightening rise ([SH87], [AR88]). Supernova neutrino experiments can provide an "alert" to the astronomical observatories, allowing the optical detection of the first stages of the supernova explosion.

The first identification of a $\bar{\nu}_e$ burst candidate is based on a statistical signature: a sudden and short increase in the counting rate of the whole scintillation counter system. This signature is preliminary to a more careful event by event analysis, which utilizes the recorded full event information. One must check if the energy distribution and the time structure of the events adequately match the expected features of a genuine neutrino burst.

[§]In practice, a stellar collapse identification in the Large Magellanic Cloud on a pure statistical basis would be possible only in Super Kamiokande.

A SuperNova Monitor software (*SNM*, fig. (7.2)) filters the positron-like *single* events (following the definition of § (4.6)) using counter to counter time and space correlations and the streamer tube information for rejecting penetrating μ's.

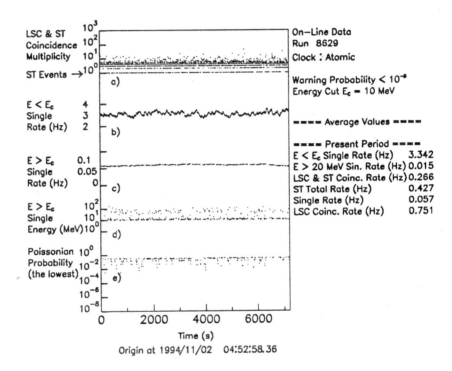

Figure 7.2: The online display of the supernova monitor program *SNM*.

This software works as a low-priority subprocess of the normal acquisition (like *LSCM*), with an event recording efficiency $\epsilon > 99\%$ at the usual acquisition rate. One might worry that in case of a very intense neutrino burst (~ 1000 events in a few seconds) the Ethernet load could be so heavy and the computer dead-time so high to significantly reduce the event recording efficiency of the spy job and, maybe, preclude the neutrino burst identification. Specific tests, performed by using a high-frequency pulser, showed that the lost event fraction is $\approx 2\%$ at a rate $R = 20$ Hz, 10% at $R = 50$ Hz and 35% at $R = 100$ Hz; these event losses are fully tolerable and are compatible with the identification of an intense $\bar{\nu}_e$ burst.*

---

*It's important to remind that this event loss refers only to the spy-job, not to the normal

The *SNM* program performs also a fast on-line energy and position reconstruction and computes the *single* event running rate (above and below a preset software energy cut E_c), the streamer tube trigger rate, the LSC and LSC & ST *coincidence* rate and the *coincidence* multiplicity and displays them as a function of time, as shown in fig. (7.2); the running rates are calculated using the last recorded 150 events, as by the *LSCM* program. The time averaged rate measured every 2 hours is recorded in a data file (*SNMGO.RAT*) and is shown on the graphical display.

A further important information provided by *SNM* (last scale in fig. (7.2)) is the lowest Poissonian probability of the event clusters. Each time a new event occurs, the event information is stored in a 4^m buffer. Such buffer is necessary, because the events are recorded and sent to the main computer by the three independent μVaxII's; so, event buffers coming from different μVaxII's can be received by the main computer in an uncorrect temporal sequence and a time window is needed to sort them out. After time ordering, the events are stored in a periodically overwritten circular buffer and an event cluster search is performed, using time intervals from 62.5 ms to 32 s in geometrical progression. For each possible cluster, the Poissonian probability of such a cluster is computed, using the last recorded background running rate; the lowest of these probabilities is displayed [MA92a].

If the software detects a neutrino burst candidate, i.e. an event cluster whose probability is lower than a preset bound (for instance 10^{-5}), an alert is immediately generated via computer nets and phone lines and a cellular phone is repeatedly called, 10 times at 2^m intervals. The "supernova experts" on shift can verify the signal credibility within 10 minutes after the alert, by using the same telephone-computer network. A schematic picture of this supernova "Early Watch" system is shown in fig. (7.3): when the *SNM* program (running on the VAXStation *WSGS04* at Gran Sasso) detects a $\bar{\nu}_e$ burst candidate, the Gran Sasso modem is activated and the cellular phone is called. The supernova expert uses the phone and a Powerbook Macintosh (equipped with an internal modem) to log-on *WSGS04* and analyze the cluster events.

Immediately after the alert the *SNM* program stores all the neutrino burst data in a dedicated DST file, produces an energy-time graphical display of the events in the burst and sends to the supernova people, via electronic mail, a summary of the cluster characteristics. An example of this summary is shown below.

```
*****************************************
WARNINGS FOR CLUSTER(S) DURING RUN007871
*****************************************
08/07/1994-18:05:27.20 PROBABILITY 7.795E-06; RATE(Hz) = 4.347E-02
WARNING! BOX 5C15: BURST DURATION 16 sec; MULTIPLICITY 8; E = 10 MeV
```

acquisition; the acquisition dead time is completely negligible: $t_{dead} < 0.1\,\% \, t_{tot}$ also at an acquisition rate $R \approx 100$ Hz.

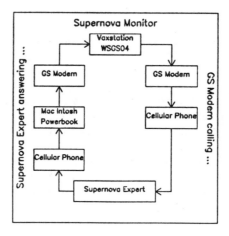

Figure 7.3: Schematic picture of the supernova "Early Watch" system

```
FULL CLUSTER INFORMATION: U.T.,BOX,DELAY(s),E(MeV)
    1)  08/07/1994-18:05:14.18   6W01(381)    0.0000    14.
    2)  08/07/1994-18:05:17.31   6B04( 84)    3.1289    11.
    3)  08/07/1994-18:05:22.21   5B11( 75)    8.0273    11.
    4)  08/07/1994-18:05:24.17   5B14( 78)    9.9961    11.
    5)  08/07/1994-18:05:24.46   6E05(485)   10.2852    12.
    6)  08/07/1994-18:05:26.10   5T17(479)   11.9258    11.
    7)  08/07/1994-18:05:27.02   5T01(265)   12.7813    11.
    8)  08/07/1994-18:05:27.20   5C15(179)   13.0273    10.
```

The summary reports the run number, the burst time, the lowest Poissonian probability, the last recorded background running rate, the last counter hit during the cluster, the neutrino burst duration and multiplicity (number of events within the burst), the energy of the last event and a complete list of the Universal Time, hit counter, delay relative to the first and energy of all the cluster events.

Finally, the "Early Watch" system is put back into normal working conditions, ready for new neutrino burst candidates.

A $\bar{\nu}_e$ burst is identified as a cluster that exceeds the probability cut and has the qualitative characteristics expected for a true stellar collapse signal: an about thermal energy spectrum, peaked at $\sim$ 12 MeV, an event temporal distribution exhibiting a sharp rise followed by a slower decrease, an event spatial distribution uniform along the apparatus etc. Some abnormal working conditions can cause false clusters, but they can be recognized by the anomalous behaviours of the monitored quantities, immediately before the burst time. Suppose for instance that the streamer tube system is suddenly misbehaving. A cluster can occur because the μ's are not well rejected and some of them are wrongly classified as *single* events. Such a cluster is identified by looking at the *coincidence* multiplicity as a function of time; it will show a systematic decreasing trend, accompanied by a systematic increasing trend of the *single* running rate; the *single* event mean energy will also increase. Similar effects are expected if some *PHRASE* modules are no longer synchronized with the others, since the μ signal identification is based on the measurement of the time correlation between events recorded by different counters.

7.3   Data automatic processing

The supernova monitor is an on-line, real time process, dedicated to the search for neutrino burst candidates; however, being a spy process, it can miss some events because of computer high dead-times or Ethernet heavy loads. A fast (off-line), 100 % efficiency, event analysis is needed to process the data safely, put them in a compact and easy-to-handle structure (*MINI.DST*) and check the monitor alarms. A permanently active batch program, based on the same algorithms used in *SNM*, checks, at 2 hour intervals, the file status and immediately processes all available data, searching for event bursts; much information about the run is stored in a text file (start and end time, error conditions detected, scintillation counter and streamer tube trigger rate every 2 hours, dead and live time, total number of events, trigger rate for each counter ...) and, for each cluster found, a mail message (similar to that previously shown) is sent to the supernova people. The text file contains also all the information related to the cluster characteristics. The *PHRASE* data are saved in a *MINI.DST* file, which contains, for each event, the Universal Time (given by the atomic clock), the hit counter, the longitudinal position, the time delay relative to the previous event, the energy and a numerical flag which distinguishes between primary ("1") and secondary ("2") events. This structure reduces the disk occupancy by more than a factor 6 if compared with the raw data: a *MINI.DST* file containing $\sim$ 90000 *PHRASE* events takes up $\sim$ 7500 VAX blocks ($\sim$ 3.5 *Mb*).

7.4   Search for event clusters and data analysis

The stellar gravitational collapse trigger is active since October 1989. A short (3 months) test run took place during the Spring 1989. The apparatus was completed in the Spring 1994 and the scintillation counter electronic equipment during the Summer of the same year. The detector active mass continuously increased and the sensitivity to a far stellar collapse improved, reaching the whole Galaxy in the Spring 1992, when all the lower part SM's were put in acquisition.

The supernova monitor is in operation since 1991; it was also improved in time reconstruction algorithms, ability in detecting hardware problems, cluster information storage etc. Up to January 1995 some hundred false event clusters were recorded and studied, most of them due to apparatus abnormal working conditions and the remaining part to background statistical fluctuations. No real supernova signal was observed.

In the search for neutrino burst we distinguish one on-line and one off-line section. The on-line section is based on the *SNM* software and is the real-time search; the off-line section is based on a data analysis which compares the observed event clusters with that predicted by the Poisson statistics. The off-line analysis gives information on the background and is a quality check for the on-line search: each event cluster having a probability lower than the warning limit, observed in the off-line analysis should have been seen by *SNM*.

Here we discuss the results of the search for neutrino bursts from stellar gravitational collapses during the period July, 27, 1994 (RUN007994) - January, 31, 1995 (RUN009232).

7.4.1   Event cluster off-line analysis

The detector active mass as a function of time during this period is shown in fig. (7.4); its average value was $\bar{M} \approx 460$ tonn.

In computing the detector active mass and up-time fraction and the distributions which will be shown in the following figures ((7.7), (7.8), ... (7.15)) some software selections were introduced: 57 runs were excluded for various kinds of anomalies (unfiltered calibration pulses, time reconstruction problems, atomic clock errors ....) and the events collected by firing or not-synchronized counters were rejected. Such selections produced a $\sim 0.4\%$ effect on the detector active mass and a $\sim 5\%$ effect on the detector up-time fraction.

The detector active mass determines the background rate and therefore the minimum number of events in a cluster with a Poissonian probability lower than the 10^{-5} preset bound. (Properly speaking the background rate also depends on the μ rejection efficiency; this is optimized when all the SM's are active and read by the data acquisition; otherwise, a μ traversing one active and an out-of-service SM can be uncorrectly flagged as a *single* event. The use of the streamer tube system reduces this effect, making it a second-order one.)

The whole MACRO running time can be divided into three sub-periods:

- from RUN007994 to RUN008244, with (almost always) 4 on-line SM's and a *single* event average trigger rate $R \simeq 41$ mHz above 10 MeV

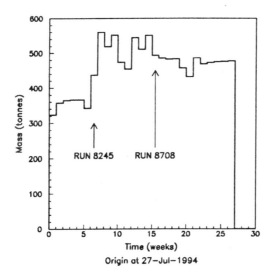

Figure 7.4: Detector active mass during the period July, 27, 1994 - January, 31, 1995 vs time.

(Period I); the SM 3 and 4 were excluded from the acquisition because of PMT gain setting operations;

- from RUN008245 to RUN008707, with (almost always) 6 on-line SM's and a *single* event average trigger rate $R \simeq 62$ mHz above 10 MeV (Period II);

- from RUN008708 to RUN009232, with (almost always) 5 on-line SM's and a *single* event average trigger rate $R \simeq 54$ mHz above 10 MeV (Period III); the SM 2 was excluded from the acquisition because of problems on some power supplies.

The 10 MeV software energy cut was chosen to improve the signal to noise ratio and reduce the fake cluster probability, as previously explained. The analysis can be easily repeated for different energy cuts and burst durations. In case of interesting burst candidates the data are indeed processed using many energy threshold values (for instance 7, 8, 10 and 12 MeV, see [BA91b]). In table (7.1) the minimum number of events N which produces a supernova burst candidate in the three periods is reported, together with the corresponding Poissonian probability P (to be multiplied by 10^{-6}); the cluster durations are selected according to the *SNM* clusterization procedure.

Table 7.1: Minimum number of events N which produces a supernova burst candidate for 6 different cluster durations in the three sub-periods and corresponding Poissonian probability P (to be multiplied by 10^{-6}).

Period	Cluster Duration (s)										
	1		2		4		8		16		32
	N	P	N	P	N	P	N	P	N	P	N	P
I	5	0.11	5	1.7	6	0.84	7	1.2	8	5.4	10	8.5
II	5	0.58	5	8.7	6	6.1	8	0.89	9	8.6	12	6.4
III	5	0.31	5	5.1	6	3.1	7	5.9	9	3.2	12	1.8

The apparatus up-time fraction during the analysis period is shown in fig. (7.5).

Since supernovæ are very rare events, the minimization of the dead-time is a fundamental goal for a stellar collapse experiment. The mean dead-time fraction during the analysis period (including software cuts) was $\approx 11.5\%$; the dead-time fraction was higher during the detector building. The present dead-time fraction is $\approx 3\%$ and is due primarily to detector maintenance and power failures; a $\sim 1^{m}$ dead time is always present at each run start, when the main computer performs the acquisition system initialization procedures.

The product of the active mass and of the up-time fraction during the analysis period is shown in fig. (7.6); this product can be regarded as a measurement of the detector "mean efficiency" in the search for supernovæ.

Figs. (7.7), (7.8) and (7.9) show the expected and the measured number of clusters vs their multiplicity ("cluster cumulative distributions") in time intervals of 2, 4, 8, 16 and 32 s for period I; figs. (7.10), (7.11) and (7.12) show the same distributions for period II and figs. (7.13), (7.14) and (7.15) for period III.

The multiplicity is defined as the number of primary events within the selected time interval. The expected number of events is computed using the Poisson statistics and the measured trigger rate during each run; the alert limits (see table (7.1)) are indicated by arrows. Note that in these plots the first event of each interval is not included; therefore the abscissa is shifted by one in respect of the table (7.1) data.

There is a general good agreement between the expected and measured distributions for each period and cluster duration. Some systematic effects are present in the long-time (16 s and 32 s) distributions; they are mainly due to sudden changes of the rejection efficiency during the run, especially by the streamer tube system. If the streamer tubes are not used, a better agreement is obtained, as shown in fig. (7.16); this distribution is the same of fig. (7.15). Note that the points corresponding to multiplicities 9 and 10 are

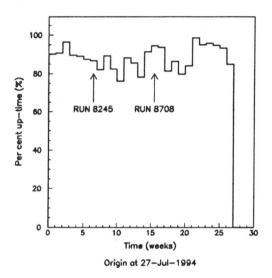

Figure 7.5: Detector up-time fraction during the period July, 27, 1994 - January, 31, 1995 vs time.

closer to the expectations than in fig. (7.15). The *single* event rate increases from ≈ 54 mHz to ≈ 60 mHz if the streamer tubes are not used.

Looking at table (7.1) and figs. (7.7), ... (7.15) one can observe that 1 event cluster, attributed to a background statistical fluctuation with a Poissonian probability less than the alert level, was recorded during Period I, 8 were recorded during Period II and 4 during Period III. Six of the clusters recorded during Period II did not cause the *SNM* software to produce an alert, because the *single* event rate at the cluster time was a bit higher than the period average value; the real probability of such clusters was therefore larger than the alert limit. Three of the clusters recorded during Period III form a unique burst of 7 events within 8 s. Then, only 5 clusters with a Poissonian probability $P < 10^{-5}$ survive: 1 during Period I (10 events within 32 s, run 8196), 2 during Period II (13 events within 32 s, run 8354 and 9 events within 16 s, run 8659; the second contains also a sub-cluster of 5 events within 2 s) and 2 during Period III (5 events within 2 s, run 9007 and 7 events within 8 s, run 8709).

Now we can check if the supernova monitor is correctly working by searching for the surviving clusters within the *SNM* alert messages; we must also provide a good explanation for each other cluster candidate detected by *SNM*.

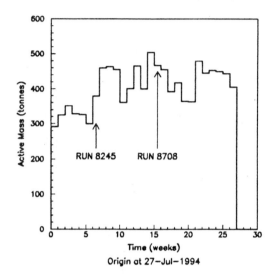

Figure 7.6: Detector active mass during the period July, 27, 1994 - January, 31, 1995 vs time, multiplied by the up-time fraction.

7.4.2 Discussion of event clusters recorded by *SNM*: a) Clusters due to apparatus anomalies

The supernova monitor recorded 53 neutrino burst candidates from July, the 27th 1994 to January, the 31st 1995.

- Six of them were due to one or few firing counters; they were identified because most events were recorded in the same box and because of their very short duration (usually < 2 s). In fig. (7.17) the ratio between the number of hit counters and the number of events in a cluster vs the cluster duration is shown; the full circles represent the "genuine" (see later) clusters and the open circles the clusters attributed to firing counters. The short clusters are strongly correlated with the low ratio values, as expected when very few counters are involved in the burst.

  The firing counters are identified by their abnormal trigger rate; these counters are then rejected from the off-line analysis and the corresponding clusters are not present in the cumulative distributions (figs. (7.7), ... (7.15)).

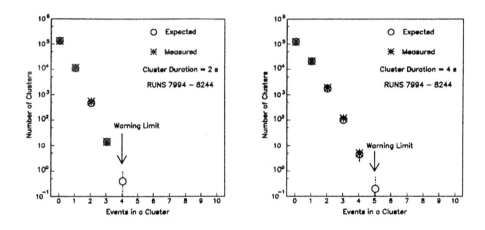

Figure 7.7: Number of event clusters vs cluster multiplicity for time intervals of 2 and 4 s during Period I.

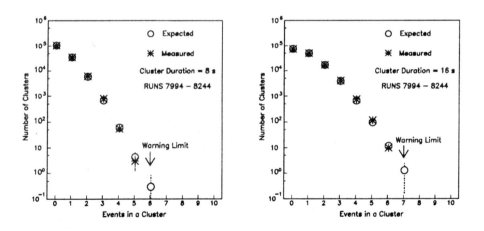

Figure 7.8: Number of event clusters vs cluster multiplicity for time intervals of 8 and 16 s during Period I.

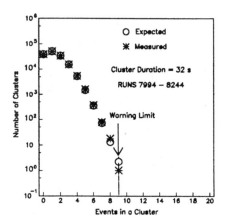

Figure 7.9: Number of event clusters vs cluster multiplicity for a time interval of 32 s during Period I.

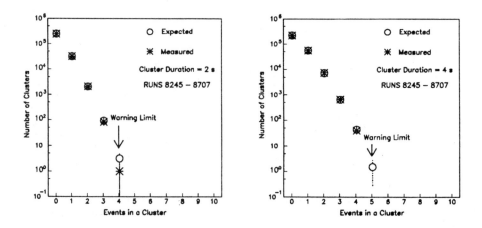

Figure 7.10: Number of event clusters vs cluster multiplicity for time intervals of 2 and 4 s during Period II.

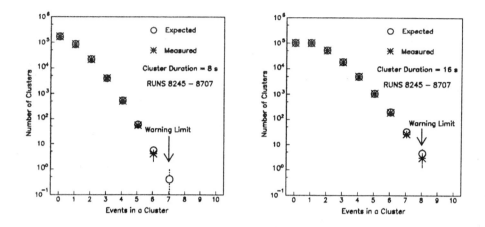

Figure 7.11: Number of event clusters vs cluster multiplicity for time intervals of 8 and 16 s during Period II.

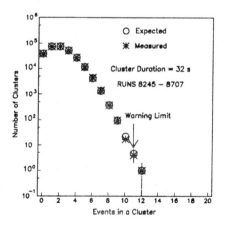

Figure 7.12: Number of event clusters vs cluster multiplicity for a time interval of 32 s during Period II.

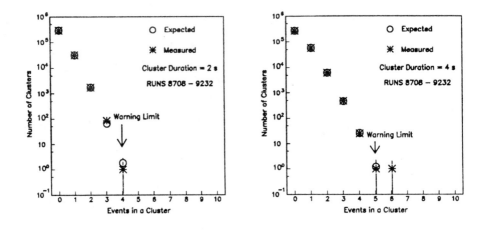

Figure 7.13: Number of event clusters vs cluster multiplicity for time intervals of 2 and 4 s during Period III.

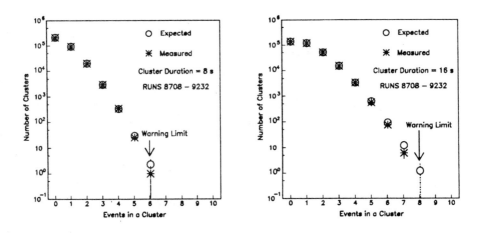

Figure 7.14: Number of event clusters vs cluster multiplicity for time intervals of 8 and 16 s during Period III.

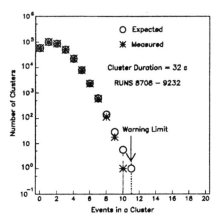

Figure 7.15: Number of event clusters vs cluster multiplicity for a time interval of 32 s during Period III.

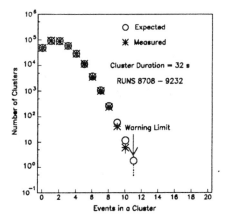

Figure 7.16: Number of event clusters vs cluster multiplicity for a time interval of 32 s during Period III; streamer tube signals were not used.

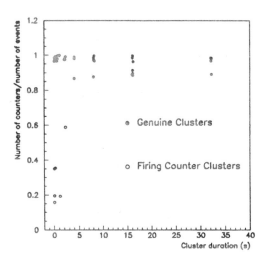

Figure 7.17: Number of hit counters divided by the number of events in the cluster vs cluster duration.

Seven clusters were due to an erroneous running rate used in calculating the Poissonian probabilities; all these clusters occurred during the maintenance days. When the maintenance operations are carried out, at least 2 supermodules are usually excluded from the acquisition; after 2 running hours, the *SNMGO.RAT* file is updated with the *single* event measured rate, which refers to a smaller number of SM's than the usual one. When all the supermodules are put back in acquisition, the Poissonian probabilities are underestimated during the first 2 running hours, because a too low rate is used. A relatively small background fluctuation can, in this case, cause a spurious cluster; this interpretation is strengthened from the fact that all these clusters occurred within 2 hours after the run start. In fig. (7.18) the ratio between the last recorded trigger rate ($E > 10$ MeV) and the number of active counters vs the time difference between the cluster occurrence and the run start is shown. The clusters having rate/N_{coun} ratios < 0.1 mHz (open circles) are concentrated within the first 2 hours after the run start, while the clusters having rate/N_{coun} $\gtrsim$ 0.11 mHz (full circles) are uniformly time-distributed.

Such clusters disappear in the off-line analysis, because the initial running rate is computed taking into account the active SM configuration.

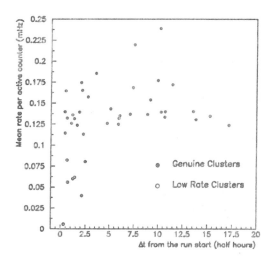

Figure 7.18: Mean trigger rate (above 10 MeV) for active counter vs time difference between the cluster and the run start.

- Six clusters were due to a large number of muons, wrongly classified as *single* events. They happened when some *PHRASE* modules were not synchronized with the others or the streamer tube trigger was not working correctly. These clusters were recognized since the event energy distributions were peaked at the expected muon energy loss and the *coincidence* and *single* trigger rates were anomalous: too low the former, too high the latter. Three of them occurred some minutes after the run start, because of problems in sending the common synchronization signal to the *PHRASE* modules; 2 others occurred within 1 hour after a strong power flicker, recorded by the power line monitor. One example: the output of a program searching for the disturbances recorded before a burst is shown

```
>>> Recorded disturbances:

    TB2 PH-B DECREASE    361 VAC         04-Jan-95 10:57:48
    TB2 PH-B INCREASE    365 VAC         04-Jan-95 10:57:48
    TB2 PH-A INCREASE    370 VAC         04-Jan-95 10:57:48
    TB2 PH-C INCREASE    371 VAC         04-Jan-95 10:57:43
```

```
TB2 PH-B INCREASE    370 VAC           04-Jan-95 10:57:48
TB2 PH-A INCREASE    374 VAC           04-Jan-95 10:57:48
TB2 PH-A DECREASE    370 VAC           04-Jan-95 10:57:48
TB2 PH-B DECREASE    365 VAC           04-Jan-95 11:05:21
TB2 PH-A DECREASE    365 VAC           04-Jan-95 11:05:21
TB2 PH-C DECREASE    366 VAC           04-Jan-95 11:05:21
TB2 PH-C DECREASE    362 VAC           04-Jan-95 11:05:22
TB2 PH-A INCREASE    371 VAC           04-Jan-95 11:05:22
TB2 PH-B DECREASE    369 VAC           04-Jan-95 11:26:54

>>> Run 9036; Cluster time:            04-Jan-95 11:44:35
```

In fig. (7.19) the event energy distributions for clusters due to a synchronization loss is compared with a (normalized) muon energy loss distribution; they are similar, as expected.

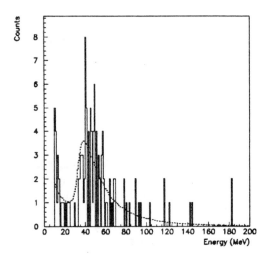

Figure 7.19: Energy distribution of the events forming the clusters due to a
synchronization loss compared with the (normalized) energy loss distribution
of the muon events.

Such clusters also disappear in the off-line analysis, because the not-
synchronized counters are rejected after being identified by periodic check-

s of internal consistency between the atomic clock and *PHRASE* event measured times.

7.4.3   Discussion of event clusters recorded by *SNM*: b) Clusters due to background fluctuations

We are then left with 34 "genuine" clusters, due to background statistical fluctuations. In fig. (7.20) the energy distribution of the events forming genuine clusters is shown (in semilogarithmic scale) and compared with the *single* event normalized energy distribution. The two energy distributions are similar and the cluster event energy spectrum is peaked close to the software energy threshold (10 MeV).

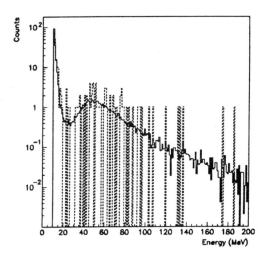

Figure 7.20: Energy distribution of the events forming the clusters due to background statistical fluctuation, compared with the normalized *single* event energy distribution.

After a more accurate energy calibration, 26 clusters disappear. Five other clusters are formed by only 3 events in a very short time interval (62.5 ms); in each cluster, at least one event was recognized as a muon because of its large energy loss.

The three surviving clusters correspond to the low-probability bursts present in the cumulative distributions during the runs 8196, 8659 and 9007. The clusters during the runs 8354 and 8709 occurred immediately before the run end; few minutes before the cluster, an error condition (a streamer tube trigger instability) was detected by the supernova monitor. Such error conditions disable the alert section of the *SNM* software; 2 and 3 events probably due to not-recognized muons are present in the clusters 8709 and 8354.

The conclusion is that the supernova monitor correctly detected all the true low-probability bursts, that are reported here:

```
*****************************************
WARNINGS FOR CLUSTER(S) DURING RUN008196
*****************************************
05/09/1994-12:08:11.85 PROBABILITY 8.727E-06; RATE(Hz) = 4.111E-02
WARNING! BOX 6E06: BURST DURATION 32 sec; MULTIPLICITY 10; E = 11 MeV
FULL CLUSTER INFORMATION: U.T.,BOX,DELAY(s),E(MeV)
     1)   05/09/1994-12:07:42.99   6E11(491)    0.0000    16.
     2)   05/09/1994-12:07:44.03   6B01( 81)    1.0430    14.
     3)   05/09/1994-12:07:50.57   6T17(495)    7.5820    65.
     4)   05/09/1994-12:07:53.93   1E12(412)   10.9492    13.
     5)   05/09/1994-12:07:57.49   2T03(219)   14.5039    10.
     6)   05/09/1994-12:07:59.74   1T03(203)   16.7539    49.
     7)   05/09/1994-12:08:00.16   1E12(412)   17.1797    11.
     8)   05/09/1994-12:08:05.33   1N07(507)   22.3438    14.
     9)   05/09/1994-12:08:07.76   6B15( 95)   24.7773    68.
    10)   05/09/1994-12:08:11.85   6E06(486)   28.8672    11.


*****************************************
WARNINGS FOR CLUSTER(S) DURING RUN008659
*****************************************
06/11/1994-12:06:57.07 PROBABILITY 8.448E-06; RATE(Hz) = 6.153E-02
WARNING! BOX 2C09: BURST DURATION  2 sec; MULTIPLICITY 5; E = 10 MeV
06/11/1994-12:07:10.56 PROBABILITY 8.175E-06; RATE(Hz) = 6.153E-02
WARNING! BOX 1W07: BURST DURATION 16 sec; MULTIPLICITY 9; E = 13 MeV
FULL CLUSTER INFORMATION: U.T.,BOX,DELAY(s),E(MeV)
     1)   06/11/1994-12:06:55.12   5T14(278)    0.0000    93.
     2)   06/11/1994-12:06:55.26   2E14(430)    0.1367    12.
     3)   06/11/1994-12:06:55.68   4C10(158)    0.5586    11.
     4)   06/11/1994-12:06:56.62   3B16( 48)    1.5000    10.
     5)   06/11/1994-12:06:57.07   2C09(125)    1.9492    10.
     6)   06/11/1994-12:07:00.51   5B01( 65)    5.3906    11.
     7)   06/11/1994-12:07:02.20   4C10(158)    7.0781    12.
     8)   06/11/1994-12:07:03.41   1B13( 13)    8.2891    10.
     9)   06/11/1994-12:07:10.56   1W07(307)   15.4453    13.


*****************************************
```

```
WARNINGS FOR CLUSTER(S) DURING RUN009007
*****************************************
30/12/1994-14:18:31.88 PROBABILITY 4.107E-06; RATE(Hz) = 5.111E-02
WARNING! BOX 6W12: BURST DURATION  2 sec; MULTIPLICITY 5; E = 11 MeV
FULL CLUSTER INFORMATION: U.T.,BOX,DELAY(s),E(MeV)
    1)   30/12/1994-14:18:30.13   4T08(256)   0.0000   10.
    2)   30/12/1994-14:18:30.20   6T12(292)   0.0732   10.
    3)   30/12/1994-14:18:30.85   1T01(201)   0.7168   11.
    4)   30/12/1994-14:18:31.74   6T11(291)   1.6094   27.
    5)   30/12/1994-14:18:31.88   6W12(392)   1.7549   11.
```

All these bursts are easily explained as due to background statistical fluctuations. The 3rd, 6th and 9th event of the 8196 burst, the 1st of the 8659 burst and the 4th of the 9007 burst are probably due to not-recognized muons; note that the energy loss of these events is large and that they are concentrated on the less shielded T layer.

The analysis of the recorded clusters is summarized in the following table.

Table 7.2: Summary of event clusters recorded by SNM in 6 months.

All clusters	
Cluster type	Cluster number
Firing counters	6
Too low initial running rate	7
Synchronization losses	6
Background fluctuations	34
Total	53
Background fluctuation clusters	
Cluster type	Cluster number
Rejected after energy calibration	26
Cluster length < 1 s	5
Surviving	3
Total	34

Chapter 8

Conclusions

The MACRO experiment is running since 5 years; during this period it performed a complete galactic survey searching for neutrinos from stellar gravitational collapses. We did not detect any genuine stellar collapse neutrino burst.

The experiment is planned to be active up to the end of the century; being sensitive to stellar collapses anywhere in the Galaxy, it will set a limit on the rate of galactic supernovæ $R_{Gal} < 0.2$/year at 90 % C.L. in case of a null signal in 10 years of data taking. Such limit is within a factor of 2 of the more optimistic predictions (see table (1.5)) and similar to that set by the Baksan collaboration. Better limits could be obtained by linking together the null signals of different experiments running in partially overlapping periods of time (there must be always at least one on-line detector); for instance, a null signal in 50 years would set a limit: $R_{Gal} < 0.045$/year at 90 % C.L. on the galactic supernova rate, corresponding to ≤ 4.5 stellar collapses/century, a realistic result. On the contrary, in case of a positive signal, MACRO would record enough $\bar{\nu}_e$ events to allow a detailed study of the neutrino spectra and of their time development. A finite $\bar{\nu}_e$ mass, in a few eV range, would also produce detectable effects in the neutrino burst time structure. The limited sensitivity to the neutral current events will probably preclude the possibility to set significant limits on ν_μ, ν_τ masses, except in case of a nearby supernova ($D \lesssim 5$ Kpc). Some ideas on how to improve such sensitivity were discussed in the past [BA90a].

The geometry of the apparatus was studied to obtain the maximum acceptance for an isotropic flux of particles and a large area; the surface/volume ratio is therefore not the best for the stellar collapse search. The liquid scintillation counters, without external shielding, are exposed to a large natural radioactivity background; the $\gamma_{2.2}$ detection efficiency is rather low, because of the active detector segmentation. On the other hand, the MACRO geometry is advantageous from the point of view of the time of flight determination: in long counters, the difference in light transit time towards the ends can be accurately measured, providing a good resolution in the position and energy reconstruction. A segmented detector is also well compatible with the tracking systems, a helpful tool for the muon background rejection.

The liquid scintillation counter experiments (like MACRO) are competitive with the water Čerenkov's. MACRO is presently one of the largest supernova liquid scintillator detector in the world; however, its active mass is a factor ≈ 50 smaller than the fiducial volume of Super Kamiokande, which will be on-line soon. We could not compete with such a larger detector if the unique comparative criterion were the number of expected events; the main strength of our experiment is the high quality of the data; our events are clean and easy to reconstruct.

Nobody knows how much time we have to wait for the next galactic stellar collapse; maybe present detectors will never detect a supernova neutrino burst, but next generation detectors will probably do. The opportunities opened by this research field are of interest for fundamental physics and astrophysics and therefore the planned long-time sky survey is a good investment for the scientific community.

Supernova neutrino experimenters look sometimes like the soldiers of an italian novel, *"Il Deserto dei Tartari"* (*"The Tartar Desert"*) by Dino Buzzati. They live for a long time in an isolated outpost in the middle of a desert, in a painful waiting for their enemies (the Tartars) to attack; and when, after years and years, the Tartars attack, the outpost is empty, because the soldiers have left it before. We obviously hope that a supernova will explode in the near future, observed by a large number of neutrino detectors; and while waiting for this astronomical event, we have the consolation of the popular wisdom: *"Patience is a virtue; possess it, if you can ..."* (english proverb) (and meanwhile write a thesis work).

Acknowledgements

A "Ph.D." thesis (expecially if the author has been working since many years in a hundred-people experiment) is never totally the work of the one who writes it; most physics results and experimental techniques discussed in this thesis are the results of the long-time work of a research team. Therefore, I am very grateful to all members of the MACRO Pisa group for their continuous support.

I'm particularly grateful to my thesis advisor, prof. C. Bemporad, for his guidance during my scientific activity and for the knowledge and the experience that he tried to pass on to me.

In the formation of a young researcher, frequent discussions with more expert people and a continuous criticism by them are fundamental. Therefore, I want to thank my "senior" colleagues, prof. R. Pazzi and G. Giannini, dr. A. Baldini and M. Grassi for their help in understanding and solving problems and stimulating careful analyses of at first sight obvious questions. I also thank my "younger" colleagues, dr. D. Nicolò and dr. G. Pieri, for their encouragement and help; likewise, I want to thank the various technicians of the MACRO Pisa group, D. Bacci, D. Picchi, M. Terachi and S. Stalio.

Sometimes this works looked me like a Sisifo stone, because I got the feeling that it would never be finished. Many INFN-Pisa young colleagues were important in comforting me when this feeling was particularly strong; I could cite many names of these friends, but I prefer to do not: if I would forget somebody, I should make an unpardonable error.

I am grateful to the National Institute for Nuclear Physics (Pisa division) for the hospitality, the technical and financial support and the use of computing resources during my thesis work; for the same reasons I am grateful to the Scuola Normale Superiore and the Gran Sasso National Laboratory.

Finally, I would like to thank my family and expecially my parents, who encouraged and sustained me when I was dejected; they showed me that, really, the life of a physicist has various advantages and many interesting aspects. This thesis is dedicated to them for their unbelievable patience.

Bibliography

[AC90]   A. Acker, S. Paksava & R. S. Raghavan, **Phys. Lett. B238**, *(1990),* *117*

[AG89a] M. Aglietta et al., **Nuovo Cimento C12**, *(1989),* *75*

[AG89b] M. Aglietta et al., **Nuovo Cimento C12**, *(1989),* *467*

[AG91]   M. Aglietta et al., **Nuovo Cimento B106**, *(1991),* *1257*

[AH91]   S. P. Ahlen et al., **MACRO Internal Memo 1002/91**, *(1991)*

[AL95]   ALEPH Collaboration, **CERN-PPE/95/03**, *(1995)*

[AR77]   W. D. Arnett, **Ap. J. 218**, *(1977),* *815*

[AR88]   W. D. Arnett, **Ap. J. 331**, *(1988),* *377*

[AR89]   W. D. Arnett et al., **Ann. Rev. Astron. Astrophys. 27**, *(1989),* *629*

[AR92a] C. Arpesella & M. Finger, Private Communication, *(1992)*

[AR92b] C. Arpesella, **LNGS Note 92/35**, *(1992)*

[AR94]   C. Arpesella et al., **LNGS Note 94/105**, *(1994)*

[AS94]   K. Assamagan et al., **Phys. Lett. B335**, *(1994),* *21*

[AT94]   E. Atzmon & S. Nussinov, **Phys. Lett. B328**, *(1994),* *103*

[BA80]   J. N. Bahcall & M. Soneira, **Ap. J. Supp.,** *44, (1980),* *73*

[BA82]   Baksan Collaboration, **Lett. Nuovo Cimento 35**, *(1982),* *413*

[BA83]   J. N. Bahcall & T. N. Piran, **Ap. J. Lett. L77**, *(1983),* *267*

[BA85]   G. Battistoni, **Nucl. Inst. and Meth. A235**, *(1985),* *91*

[BA86]   G. Battistoni, **Nucl. Inst. and Meth. A247**, *(1986),* *277*

[BA87a] J. N. Bahcall, T. N. Piran, W. H. Press & D. N. Spergel, **Nature 327**, (1987), 682

[BA87b] Baksan Collaboration, **JETP Lett. 45**, (1987), 589

[BA87c] G. Barbiellini & G. Cocconi, **Nature 329**, (1987), 21

[BA88a] J. N. Bahcall, K. Kubodera & S. Kunozawa, **Phys. Rev., D38**, (1988), 1030

[BA88b] R. Barbieri & R. Mohapatra, **Phys. Rev. Lett. 61**, (1988), 27

[BA88c] Baksan Collaboration, **Phys. Lett. B205**, (1988), 209

[BA89a] J. N. Bahcall, "Neutrino Astrophysics", Cambridge University Press, (1989)

[BA89b] J. N. Bahcall & W. Haxton, **Phys. Rev. D40**, (1989), 3211

[BA89c] G. Battistoni, **Nucl. Inst. and Meth. A279**, (1989), 137

[BA90a] A. Baldini et al., **MACRO Internal Memo 17/90**, (1990)

[BA90b] A. Baldini et al., **MACRO Internal Memo 18/90**, (1990)

[BA91a] A. Baldini et al., **Nucl. Inst. and Meth. A305**, (1991), 475

[BA91b] A. Baldini et al., **MACRO Internal Memo 13/91**, (1991)

[BA91c] A. Baldini et al., **MACRO Internal Memo 20/91**, (1991)

[BA91d] A. Baldini et al., **MACRO Internal Memo 27/91**, (1991)

[BA91e] A. Baldini et al., **MACRO Internal Memo 28/91**, (1991)

[BA92a] A. Baldini et al., **MACRO Internal Memo 10/92**, (1992)

[BA93a] A. Baldini et al., **MACRO Internal Memo 8/93**, (1993)

[BA93b] J. N. Bahcall, **IASSNS-AST 93/69**, (1993)

[BA94]   Baksan Collaboration, **Nucl. Phys. (Proc. Suppl.) B35**, (1994), 270

[BA95]   A. Baldini et al., **MACRO Internal Memo 3/95**, (1995)

[BE85]   E. Bellotti, **INFN/TC Report, 85/19**, (1985)

[BE88]   E. Bellotti, **Nucl. Inst. and Meth. A264**, (1988), 1

[BE89]   P. Belli et al., **Il Nuovo Cim. A101**, (1989), 959

[BE90a] S. Bermon et al., **Phys. Rev. Lett. 64**, (1990), 839

[BE90b] H. A. Bethe, **Rev. Mod. Phys. 62**, (1990), 801

[BE91] H. A. Bethe & J. N. Bahcall, **Phys. Rev. D44**, (1991), 2962

[BE92] E. W. Beier et al., **Phys. Lett. B283**, (1992), 446

[BE93] V. Berezinsky, **LNGS Note 93/86**, (1993)

[BE94] A. I. Belesev et al., preprint **INR 862/94**, (1994)

[BI51] J. L. Birks, **Proc. Phys. Soc. A64**, (1951), 874

[BI78] S. M. Bilenky & B. M. Pontecorvo, **Phys. Rep. 41**, (1978), 225

[BL88] S. A. Bludman & P. J. Schinder, **Ap. J. 326**, (1988), 265

[BL92] S. A. Bludman, D. C. Kennedy & P. G. Langacker, **Phys. Rev. D45**, (1992), 1810

[BN64] Brookhaven National Laboratory 325, "Neutron Cross Sections", 2nd Edition, (1964)

[BO80] J. R. Bond et al. **Phys. Rev. Lett. 45**, (1980), 61

[BO84] F. Bourgeois, **Nucl. Inst. and Meth. 219**, (1984), 153

[BÖ92] Felix Böhm & Peter Vogel, "Physics of Massive Neutrinos", 2nd Edition, Cambridge University Press, (1992)

[BR79] F. D. Brooks, **Nucl. Inst. and Meth. 162**, (1979), 477

[BR87] S. W. Bruenn, **Phys. Rev. Lett. 59**, (1987), 938

[BR91] S. W. Bruenn & W. Haxton, **Ap. J. 376**, (1991), 678

[BU88] A. Burrows & J. M. Lattimer, **Ap. J. 307**, (1988), 265

[BU90a] K. N. Buckland et al., **Phys. Rev. D41**, (1990), 2726

[BU90b] A. Burrows, **Ann. Rev. Nucl. Part. Sci. 40**, (1990), 181

[BU92] A. Burrows, D. Klein & R. Gandhi, **Phys. Rev. D45**, (1992), 3361

[BU93] A. Burrows, D. Klein & R. Gandhi, **Nucl. Phys. (Proc. Suppl.), B31**, (1993), 408

[BU94] A. Burrows in "Proceedings of the Summer Study on Nuclear and Particle Astrophysics and Cosmology in the Next Millennium", Snowmass (Colorado) 29 June-14 July 1994 (to be published)

164   Fabrizio Cei

[BU95]   A. Burrows, J. Hayes & B. Fryxell, preprint astro-ph 9506061,
          (1995)

[CA94a]   D. O. Caldwell in "Proceedings of the Neutrino '94 Conference",
          Eilat (Israel), May 29-June 3, 1994

[CA94b]   V. Castellani et al., Phys. Rev. D50, (1994), 4749

[CL90]   D. Cline, Astro. Lett. and Communications, 27, (1990), 403

[CL93]   D. Cline in "Proceedings of the Venice Workshop on Neutrino Phy-
          sics", Venice (Italy), (1993)

[CL94]   D. Cline, G. M. Fuller, W. P. Hong, B. Meyer & J. Wilson, Phys.
          Rev. D50, (1994), 720

[CO60]   S. A. Colgate & M. H. Johnson, Phys. Rev. Lett. 5, (1960), 235

[CO72]   R. Cowsik & J. Mc Celland, Phys. Rev. Lett. 29, (1972), 669

[DA89]   I. D'Antone et al., IEEE Transactions on Nuclear Science, 36,
          (1989), 1602

[DA93]   R. Davis Jr. in "Proceedings of the 23nd ICRC", Calgary (Canada),
          3, (1993), 869

[DO79]   T. W. Donnelly & R. D. Peccei, Phys. Rep. 50, (1979), 1

[DR83]   S. D. Drell et al., Phys. Rev. Lett. 50, (1983), 644

[DR87]   A. De Rújula et al., Phys. Lett. B193, (1987), 514

[EA89]   EAS-TOP Collaboration, Nucl. Inst. and Meth. A277, (1989),
          23

[FE88]   F. von Feilitzsch & L. Oberauer Phys. Lett. 200, (1988), 580

[FI85]   E. Fiorini et al., INFN Internal Note LNF-85/7(R), (1985)

[FR88]   J. Frieman, E. Haber & K. Freese, Phys. Lett. B200, (1988), 115

[FR93]   W. Frati et al., Phys. Rev. D48, (1993), 1140

[FU87]   G. M. Fuller, R. Mayle, J. R. Wilson & D. N. Schramm, Ap. J.
          322, (1987), 795

[FU88]   M. Fukugita et al., Phys. Lett. B212, (1988), 139

[FU92]   G. M. Fuller, R. Mayle, B. S. Meyer & J. R. Wilson, Ap. J. 389,
          (1992), 517

[GA90]   R. Gandhi & A. Burrows, Phys. Lett. B246, (1990), 149

[GA92a]   GALLEX Collaboration, Phys. Lett. B285, (1992), 376

[GA92b]   GALLEX Collaboration, Phys. Lett. B285, (1992), 390

[GA94a]   GALLEX Collaboration, Phys. Lett. B327, (1994), 377

[GA94b]   T. K. Gaisser in "Proceedings of the Summer Study on Nuclear and Particle Astrophysics and Cosmology in the Next Millennium", Snowmass (Colorado) 29 June-14 July 1994 (to be published)

[GU78]   J. E. Gunn et al., Ap. J. 223, (1978), 1015

[HA87]   W. Haxton, Phys. Rev. D36, (1987), 2283

[HA88a]   W. Haxton, Phys. Rev. C37, (1988), 2660

[HA88b]   W. Haxton, Phys. Rev. Lett. 60, (1988), 1999

[HA89]   W. Haxton et al., Phys. Rev. D40, (1989), 3211

[HA94]   F. Halzen, J. E. Jacobsen & E. Zas, Phys. Rev. D49, (1994), 1758

[HO74]   G. 't Hooft, Nucl. Phys. B79, (1974), 276

[HU90]   M. E. Huber, Phys. Rev. Lett. 64, (1990), 835

[IA83]   E. Iarocci, Nucl. Inst. and Meth. A217, (1983), 30

[IC94]   ICARUS Collaboration, ICARUS Proposal, LNGS Note 94/99-I, (1994)

[IM87]   IMB Collaboration, Phys. Rev. Lett. 58, (1987), 1494

[IM92]   IMB Collaboration, Phys. Rev. D46, (1992), 3720

[JA93]   H. T. Janka in "Proceedings of Frontier Objects in Astrophysics and Particle Physics", Vulcano (Italy), (1993)

[JA94]   H. T. Janka in "Proceedings of the Neutrino '94 Conference", Eilat (Israel), May 29-June 3, 1994

[KA87]   Kamiokande Collaboration, Phys. Rev. Lett. 58, (1987), 1490

[KA88]   Kamiokande Collaboration, Phys. Rev. Lett. 61, (1988), 385

[KA90]   Kamiokande Collaboration, Phys. Rev. Lett. 65, (1990), 1297

[KA91a]   Kamiokande Collaboration, Phys. Rev. D44, (1991), 2241

[KA91b] *Kamiokande Collaboration*, **ICRR Report 239-91-8**, *(1991)*

[KA92] *Kamiokande Collaboration*, **Phys. Lett. B280**, *(1992), 146*

[KA93] *KARMEN Collaboration*, **Nucl. Phys. A553**, *(1993), 931c*

[KA94] *KARMEN Collaboration in "Proceedings of the Summer Study on Nuclear and Particle Astrophysics and Cosmology in the Next Millennium", Snowmass (Colorado) 29 June-14 July 1994 (to be published)*

[KE86] *F. J. Kerr & D. Lynden-Bell*, **Mon. Not. R. Astron. Soc. 221**, *(1986), 1023*

[KO92] *M. Koshiba*, **Phys. Rep. 220**, *(1992), 229*

[KR84] *L. M. Krauss, S. L. Glashow & D. N. Schramm*, **Nature 310**, *(1984), 191*

[KR88] *L. M. Krauss & S. Tremaine*, **Phys. Rev. Lett. 60**, *(1988), 176*

[LA94] *"LABVIEW User Manual for Macintosh", edited by National Instrument Corporation, (1994)*

[LE78] *C. M. Lederer, J. M. Hollander & I. Perlman, "Table of Isotopes", 7th Edition, (1978)*

[LE92] *W. R. Leo, "Techniques for Nuclear and Particle Physics Experiments", Ed. Springer-Verlag (1992)*

[LI88] *G. Liu et al.*, **MACRO Internal Memo 1008/88**, *(1988)*

[LI93] *R. Liu*, **MACRO Internal Memo 11/93**, *(1993)*

[LO88] *J. M. Longo*, **Phys. Rev. Lett. 60**, *(1988), 173*

[LS87] *LSD Collaboration*, **Europhys. Lett. 3**, *(1987), 1315*

[LS92] *LSD Collaboration*, **Astropart. Phys. 1**, *(1992), 1*

[LV88] *LVD Collaboration*, **Nucl. Inst. and Meth. A264**, *(1988), 5*

[LV92] *LVD Collaboration*, **Nuovo Cimento, A105**, *(1992), 1793*

[LV93] *LVD Collaboration*, **Nucl. Phys. (Proc. Suppl.) B31**, *(1993), 450*

[MA84] *MACRO Collaboration*, **MACRO Proposal**, *(1984)*

[MA87a] *R. Mayle, J. R. Wilson & D. N. Schramm*, **Ap. J. 318**, *(1987), 288*

[MA87b] *R. M. Mc Naught, IAU Circular N° 4316, (1987)*

[MA88a] *MACRO Collaboration*, Nucl. Inst. and Meth. **A264**, *(1988), 18*

[MA88b] *MACRO Collaboration*, Nucl. Tracks Radiat. Meas. **15**, *(1988), 331*

[MA90] *MACRO Collaboration*, Phys. Lett. **B249**, *(1990), 149*

[MA92a] *MACRO Collaboration*, Astroparticle Phys. **1**, *(1992), 11*

[MA92b] *MACRO Collaboration*, Phys. Rev. Lett. **69**, *(1992), 1860*

[MA92c] *MACRO Collaboration*, Phys. Rev. **D46**, *(1992), 895*

[MA92d] *MACRO Collaboration*, Phys. Rev. **D46**, *(1992), 4836*

[MA93a] *MACRO Collaboration*, Nucl. Inst. and Meth. **A324**, *(1993), 337*

[MA93b] *MACRO Collaboration*, Ap. J. **412**, *(1993), 301*

[MA93c] *MACRO Collaboration in "Proceedings of the International Conference on Theoretical and Phenomenological Aspects of Underground Physics" (TAUP93), L' Aquila (Italy), 19-23 september 1993*

[MA94a] *MACRO Collaboration*, Phys. Rev. Lett. **72**, *(1994), 608*

[MA94b] *MACRO Collaboration*, **MACRO/PUB** Internal Note, **94/4**, *(1994)*

[MA95a] *MACRO Collaboration, submitted to* Phys. Lett. **B**, *(1995)*

[MA95b] *MACRO Collaboration in "Proceedings of the International Conference on Theoretical and Phenomenological Aspects of Underground Physics" (TAUP95), Toledo (Spain), 18-23 september 1995*

[MA95c] *MACRO Collaboration*, Astropart. Phys **4**, *(1995), 33-43*

[ME90] *MACRO & EAS-TOP Collaboration*, Phys. Rev. **D42**, *(1990), 1396*

[ME94] *MACRO & EAS-TOP Collaboration*, Phys. Lett. **B337**, *(1994), 376*

[MI72] *G. H. Miller et al.*, Phys. Lett. **B41**, *(1972), 50*

[MI82] *S. L. Mintz*, Phys. Rev. Lett. **C25**, *(1982), 1671*

[MI86] *S. P. Mikheyev & A. Y. Smirnov*, Nuovo Cimento **C9**, *(1986), 17*

[MI93] *L. Mikaelyan et al.*, Kurchatov Institute Report **50.05/111**, *(1993)*

[MO91]   R. N. Mohapatra & P. N. Bal, "Massive Neutrinos in Physics and Astrophysics", Ed. World Scientific, (1991)

[MU76]   N. Mukuyama et al., Nucl. Instr. and Meth. A134, (1976), 125

[MY90]   S. Myra & A. Burrows, Ap. J. 364, (1990), 222

[NA80]   D. K. Nadezhin & I. V. Otroshchenko, Astron. Zh. 57, (1980), 78

[NA87]   R. Narayan, Ap. J. 319, (1987), 162

[NA88]   M. Nakahata, Ph. D Thesis, Unniversity of Tokio, preprint UT-ICEPP-88/01, (1988)

[OR91]   S. Orito et al., Phys. Rev. Lett. 66, (1991), 1951

[PA82]   E. N. Parker, T. N. Bogdan & M. S. Turner, Phys. Rev. D26, (1982), 1296

[PI89]   T. Piran et al., Nature 340, (1989), 126

[PI90]   T. Piran in "Supernovæ", Jerusalem Winter School for Theoretical Physics, Ed. J. C. Wheleer, T. Piran & S. Weinberg, (1990), 303

[PI92]   T. Piran et al., Ap. J. 395, (1992), L83

[PI93]   T. Piran & C. S. Kochanek, Ap. J. 417, (1993), L17

[PO68]   B. M. Pontecorvo, Sov. JETP 26, (1968), 984

[PO74]   M. Polyakov, JETP Lett. 20, (1974), 194

[PR84]   J. Preskill, Ann. Rev. Nucl. Part. Sci. 34, (1984), 461

[PR86]   W. H. Press et al., "Numerical Recipes", Cambridge University Press, (1986)

[QI93]   Y. Qian, G.M. Fuller, G. J. Mathews, R. W. Mayle & J. R. Wilson, Phys. Rev. Lett. 71, (1993), 1965

[QI94]   Y. Qian & G.M. Fuller, Phys. Rev. D49, (1994), 1762

[QI95]   Y. Qian & G.M. Fuller, Phys. Rev. D51, (1995), 1479

[RA88]   G. Raffelt & D. Seckel, Phys. Rev. Lett. 60, (1988), 1793

[RA89]   K. Ratnatunga & S. Van den Bergh, Ap. J. 343, (1989), 713

[RI88]   A. Rindi, LNF Report, 88/01, (1988)

[RY92]   O. Ryazhskaya, JETP Lett. 56, (1992), 417

[RY93]   O. Ryazhskaya, V. G. Ryasny & O. Saavedra in "Proceedings of the 23nd ICRC", Calgary (Canada), 4, (1993), 480

[RY94]   O. Ryazhskaya, V. G. Ryasny & O. Saavedra, Pis'ma Zh. Eksp. Teor. Fiz. 59, (1994), 297

[SC88]   D. N. Schramm, Nucl. Phys. B3 (Proc. Suppl.), (1988), 471

[SC90]   D. N. Schramm & L. E. Brown in "Supernovæ", Jerusalem Winter School for Theoretical Physics, Ed. J. C. Wheeler, T. Piran & S. Weinberg, (1990), 261

[SC92]   R. K. Schaefer & Q. Shafi, Nature 359, (1992), 199

[SE77]   E. Segrè "Nuclei and Particles", 2nd edition, W. A. Benjamin Inc. Press, (1977)

[SH87]   T. Shigeyama, K. Nomoto, M. Hashimoto & D. Sugimoto, Nature 328, (1987), 320

[SM92]   G. F. Smoot et al., Ap. J. 396, (1992), L1

[SM94]   A. Smirnov, D. N. Spergel & J. N. Bahcall, Phys. Rev. D49, (1994), 1389

[SN88]   SNO Collaboration, Nucl. Inst. and Meth. A264, (1988), 48

[SN94]   SNO Collaboration in "Proceedings of the Summer Study on Nuclear and Particle Astrophysics and Cosmology in the Next Millennium", Snowmass (Colorado) 29 June-14 July 1994 (to be published)

[SO92]   SOUDAN 2 Collaboration, Phys. Rev. D46, (1992), 4846

[ST88]   L. Stodolsky, Phys. Lett. B201, (1988), 353

[SW94]   D. Swesty in "Proceedings of the Summer Study on Nuclear and Particle Astrophysics and Cosmology in the Next Millennium", Snowmass (Colorado) 29 June-14 July 1994 (to be published)

[TA82]   G. Tammann in "Supernovæ: A Survey of Current Research", Ed. M. J. Rees, (1982)

[TO87]   Y. Totsuka in "Proceedings of the Seventh Workshop on Grand Unification", Toyama (Japan), Ed. J. Arafune, (1987)

[TO95]   T. Totani & K. Sato, preprint UTAP - 203/95, (1995)

[VA91a] S. Van den Bergh & G. Tammann, Ann. Rev. Astron. Astrophys. 29, (1991), 363

[VA91b] *S. Van den Bergh,* **Phys. Rep. 204,** *(1991), 385*

[VA93]   *S. Van den Bergh,* **Comments Astrophys. 17,** *(1993), 125*

[WH90]   *J. C. Wheeler in "Supernovæ", Jerusalem Winter School for Theoretical Physics, Ed. J. C. Wheeler, T. Piran & S. Weinberg, (1990), 1*

[WI85]   *J. R. Wilson in "Numerical Astrophysics", Ed. Jones & Bartlett, Boston, (1985)*

[WI93]   *J. R. Wilson & R. W. Mayle,* **Phys. Rep. 227,** *(1993), 97*

[WO78]   *L. Wolfenstein,* **Phys. Rev. D17,** *(1978), 2369*

[WO86a] *S. E. Woosley & T. A. Weaver,* **Ann. Rev. Astron. Astrophys. 24,** *(1986), 205*

[WO86b] *S. E. Woosley, J. R. Wilson & R. Mayle,* **Ap. J. 302,** *(1986), 19*

Elenco delle Tesi di perfezionamento della Classe di Scienze
pubblicate dall'Anno Accademico 1992/93

HISAO FUJITA YASHIMA, *Equations de Navier–Stokes stochastiques non homogènes et applications*, 1992.

GIORGIO GAMBERINI, *The Minimal supersymmetric standard model and its phenomenological implications*, 1993.

CHIARA DE FABRITIIS, *Actions of Holomorphic Maps on Spaces of Holomorphic Functions*, 1994.

CARLO PETRONIO, *Standard Spines and 3-Manifolds*, 1995.

MARCO MANETTI, *Degenerations of Algebraic Surfaces and Applications to Moduli Problems*, 1995.

ILARIA DAMIANI, *Untwisted Affine Quantum Algebras: the Highest Coefficient of $detH_\eta$ and the Center at Odd Roots of 1*, 1995.

FABRIZIO CEI, *Search for Neutrinos from Stellar Gravitational Collapse with the MACRO Experiment at Gran Sasso*, 1996.

Pantograf s.n.c. - Via alla Stazione di Voltri, 2/A - Genova
Finito di stampare nel marzo 1996